Whole-Angle MEMS Gyroscopes

IEEE Press
445 Hoes Lane
Piscataway, NJ 08854

IEEE Press Editorial Board
Ekram Hossain, *Editor in Chief*

Jón Atli Benediktsson	Bimal Bose	David Alan Grier
Elya B. Joffe	Xiaoou Li	Peter Lian
Andreas Molisch	Saeid Nahavandi	Jeffrey Reed
Diomidis Spinellis	Sarah Spurgeon	Ahmet Murat Tekalp

Whole-Angle MEMS Gyroscopes

Challenges and Opportunities

Doruk Senkal

Andrei M. Shkel

IEEE Press Series on Sensors
Vladimir Lumelsky, Series Editor

Copyright © 2020 by The Institute of Electrical and Electronics Engineers, Inc. All rights reserved.

Published by John Wiley & Sons, Inc., Hoboken, New Jersey.
Published simultaneously in Canada.

No part of this publication may be reproduced, stored in a retrieval system, or transmitted in any form or by any means, electronic, mechanical, photocopying, recording, scanning, or otherwise, except as permitted under Section 107 or 108 of the 1976 United States Copyright Act, without either the prior written permission of the Publisher, or authorization through payment of the appropriate per-copy fee to the Copyright Clearance Center, Inc., 222 Rosewood Drive, Danvers, MA 01923, (978) 750-8400, fax (978) 750-4470, or on the web at www.copyright.com. Requests to the Publisher for permission should be addressed to the Permissions Department, John Wiley & Sons, Inc., 111 River Street, Hoboken, NJ 07030, (201) 748-6011, fax (201) 748-6008, or online at http://www.wiley.com/go/permission.

Limit of Liability/Disclaimer of Warranty: While the publisher and author have used their best efforts in preparing this book, they make no representations or warranties with respect to the accuracy or completeness of the contents of this book and specifically disclaim any implied warranties of merchantability or fitness for a particular purpose. No warranty may be created or extended by sales representatives or written sales materials. The advice and strategies contained herein may not be suitable for your situation. You should consult with a professional where appropriate. Neither the publisher nor author shall be liable for any loss of profit or any other commercial damages, including but not limited to special, incidental, consequential, or other damages.

For general information on our other products and services or for technical support, please contact our Customer Care Department within the United States at (800) 762-2974, outside the United States at (317) 572-3993 or fax (317) 572-4002.

Wiley also publishes its books in a variety of electronic formats. Some content that appears in print may not be available in electronic formats. For more information about Wiley products, visit our web site at www.wiley.com.

Library of Congress Cataloging-in-Publication Data is applied for

ISBN 9781119441885

Cover Design: Wiley
Cover Image: Courtesy of Doruk Senkal

Contents

List of Abbreviations *ix*
Preface *xi*
About the Authors *xiii*

Part I Fundamentals of Whole-Angle Gyroscopes *1*

1   **Introduction** *3*
1.1   Types of Coriolis Vibratory Gyroscopes *3*
1.1.1   Nondegenerate Mode Gyroscopes *4*
1.1.2   Degenerate Mode Gyroscopes *5*
1.2   Generalized CVG Errors *5*
1.2.1   Scale Factor Errors *7*
1.2.2   Bias Errors *7*
1.2.3   Noise Processes *7*
1.2.3.1 Allan Variance *7*
1.3   Overview *9*

2   **Dynamics** *11*
2.1   Introduction to Whole-Angle Gyroscopes *11*
2.2   Foucault Pendulum Analogy *11*
2.2.1   Damping and Q-factor *12*
2.2.1.1 Viscous Damping *13*
2.2.1.2 Anchor Losses *14*
2.2.1.3 Material Losses *15*
2.2.1.4 Surface Losses *16*
2.2.1.5 Mode Coupling Losses *16*
2.2.1.6 Additional Dissipation Mechanisms *16*
2.2.2   Principal Axes of Elasticity and Damping *16*

2.3 Canonical Variables *18*
2.4 Effect of Structural Imperfections *18*
2.5 Challenges of Whole-Angle Gyroscopes *20*

3 Control Strategies *23*
3.1 Quadrature and Coriolis Duality *23*
3.2 Rate Gyroscope Mechanization *24*
3.2.1 Open-loop Mechanization *24*
3.2.1.1 Drive Mode Oscillator *24*
3.2.1.2 Amplitude Gain Control *26*
3.2.1.3 Phase Locked Loop/Demodulation *26*
3.2.1.4 Quadrature Cancellation *26*
3.2.2 Force-to-rebalance Mechanization *27*
3.2.2.1 Force-to-rebalance Loop *27*
3.2.2.2 Quadrature Null Loop *29*
3.3 Whole-Angle Mechanization *29*
3.3.1 Control System Overview *30*
3.3.2 Amplitude Gain Control *32*
3.3.2.1 Vector Drive *32*
3.3.2.2 Parametric Drive *33*
3.3.3 Quadrature Null Loop *34*
3.3.3.1 AC Quadrature Null *34*
3.3.3.2 DC Quadrature Null *34*
3.3.4 Force-to-rebalance and Virtual Carouseling *35*
3.4 Conclusions *35*

Part II 2-D Micro-Machined Whole-Angle Gyroscope Architectures *37*

4 Overview of 2-D Micro-Machined Whole-Angle Gyroscopes *39*
4.1 2-D Micro-Machined Whole-Angle Gyroscope Architectures *39*
4.1.1 Lumped Mass Systems *39*
4.1.2 Ring/Disk Systems *40*
4.1.2.1 Ring Gyroscopes *40*
4.1.2.2 Concentric Ring Systems *41*
4.1.2.3 Disk Gyroscopes *42*
4.2 2-D Micro-Machining Processes *42*
4.2.1 Traditional Silicon MEMS Process *43*
4.2.2 Integrated MEMS/CMOS Fabrication Process *43*
4.2.3 Epitaxial Silicon Encapsulation Process *44*

5 Example 2-D Micro-Machined Whole-Angle Gyroscopes 47
5.1   A Distributed Mass MEMS Gyroscope – Toroidal Ring Gyroscope 47
5.1.1   Architecture 48
5.1.1.1 Electrode Architecture 49
5.1.2   Experimental Demonstration of the Concept 49
5.1.2.1 Fabrication 49
5.1.2.2 Experimental Setup 50
5.1.2.3 Mechanical Characterization 51
5.1.2.4 Rate Gyroscope Operation 52
5.1.2.5 Comparison of Vector Drive and Parametric Drive 53
5.2   A Lumped Mass MEMS Gyroscope – Dual Foucault Pendulum Gyroscope 54
5.2.1   Architecture 56
5.2.1.1 Electrode Architecture 57
5.2.2   Experimental Demonstration of the Concept 57
5.2.2.1 Fabrication 57
5.2.2.2 Experimental Setup 58
5.2.2.3 Mechanical Characterization 60
5.2.2.4 Rate Gyroscope Operation 60
5.2.2.5 Parameter Identification 60

Part III 3-D Micro-Machined Whole-Angle Gyroscope Architectures 65

6 Overview of 3-D Shell Implementations 67
6.1   Macro-scale Hemispherical Resonator Gyroscopes 67
6.2   3-D Micro-Shell Fabrication Processes 69
6.2.1   Bulk Micro-Machining Processes 69
6.2.2   Surface-Micro-Machined Micro-Shell Resonators 74
6.3   Transduction of 3-D Micro-Shell Resonators 79
6.3.1   Electromagnetic Excitation 79
6.3.2   Optomechanical Detection 80
6.3.3   Electrostatic Transduction 81

7 Design and Fabrication of Micro-glassblown Wineglass Resonators 87
7.1   Design of Micro-Glassblown Wineglass Resonators 88
7.1.1   Design of Micro-Wineglass Geometry 90
7.1.1.1 Analytical Solution 90
7.1.1.2 Finite Element Analysis 92

7.1.1.3 Effect of Stem Geometry on Anchor Loss *94*
7.1.2 Design for High Frequency Symmetry *96*
7.1.2.1 Frequency Symmetry Scaling Laws *97*
7.1.2.2 Stability of Micro-Glassblown Structures *101*
7.2 An Example Fabrication Process for Micro-glassblown Wineglass Resonators *102*
7.2.1 Substrate Preparation *103*
7.2.2 Wafer Bonding *103*
7.2.3 Micro-Glassblowing *104*
7.2.4 Wineglass Release *105*
7.3 Characterization of Micro-Glassblown Shells *106*
7.3.1 Surface Roughness *107*
7.3.2 Material Composition *108*

8 Transduction of Micro-Glassblown Wineglass Resonators *111*
8.1 Assembled Electrodes *111*
8.1.1 Design *111*
8.1.2 Fabrication *112*
8.1.2.1 Experimental Characterization *113*
8.2 In-plane Electrodes *115*
8.3 Fabrication *115*
8.4 Experimental Characterization *118*
8.5 Out-of-plane Electrodes *123*
8.6 Design *123*
8.7 Fabrication *126*
8.8 Experimental Characterization *129*

9 Conclusions and Future Trends *133*
9.1 Mechanical Trimming of Structural Imperfections *133*
9.2 Self-calibration *134*
9.3 Integration and Packaging *135*

References *137*

Index *149*

List of Abbreviations

Table 1 Control system abbreviations.

Symbol	Description
Quadrature	Unwanted component of oscillation that interferes with estimation of the pattern angle, manifests as a result of structural imperfections
AGC	Amplitude Gain Control, closed-loop control of drive amplitude
PLL	Phase Locked Loop, closed-loop control system that generates an AC signal with a predetermined phase offset from the resonator
FTR	Force-to-rebalance, closed-loop control system that actively drives the pattern angle to a setpoint
Quadrature null	Closed-loop control system that actively suppresses the effects of structural imperfections within the gyroscope

Table 2 Mechanical parameters of the resonator.

Symbol	Description
f	Mean frequency of the two primary modes of the resonator
τ	Mean energy decay time constant of the resonator
Q-factor	Ratio of stored energy to energy loss per vibration cycle ($Q = \tau \pi f$)
Δf	Frequency split between primary modes in Hz ($\Delta f = f_x - f_y$)
$\Delta \omega$	Frequency split between primary modes in rad/s ($\Delta \omega = \omega_x - \omega_y$)
$\Delta \tau^{-1}$	Measure of anisodamping within the gyroscope ($\Delta \tau^{-1} = \vert \tau_x^{-1} - \tau_y^{-1} \vert$)
θ_ω	Angle defining the orientation of actual versus intended axes of elasticity
θ_τ	Angle defining the orientation of actual versus intended axes of damping
(x, y, z)	Coordinate frame oriented along intended axes of symmetry x and y
$n = 2$ mode	A 4-node degenerate mode pair of a wineglass or ring/disk system
$n = 3$ mode	A 6-node degenerate mode pair of a wineglass or ring/disk system
Precession pattern	Vibration pattern formed by superposition of x and y vibratory modes, which is capable of changing its orientation (precesses) when subjected to Coriolis forces or an external forcing function
Pattern angle (θ)	Orientation of the precession pattern in degrees, which is a measure of angular rotation in a Rate Integrating Gyroscope

Preface

Coriolis Vibratory Gyroscopes (CVGs) can be divided into two broad categories based on the gyroscope's mechanical element: degenerate mode gyroscopes (type 1), which have x–y symmetry, and nondegenerate mode gyroscopes (type 2), which are designed intentionally to be asymmetric in x and y modes.

Currently, nondegenerate mode gyroscopes fulfill the needs of a variety of commercial applications, such as tilt detection, activity tracking, and gaming. However, when it comes to inertial navigation, where sensitivity and stability of the sensors are very important, commercially available MEMS sensors fall short by three orders of magnitude. Degenerate mode gyroscopes, on the other hand, offer a number of unique advantages compared to nondegenerate vibratory rate gyroscopes, including higher rate sensitivity, ability to implement whole-angle mechanization with mechanically unlimited dynamic range, exceptional scale factor stability, and a potential for self-calibration. For this reason, as the MEMS gyroscope development is reaching maturity, the Research and Development focus is shifting from high-volume production of low-cost nondegenerate mode gyroscopes to high performance degenerate mode gyroscopes. This paradigm shift in MEMS gyroscope research and development creates a need for a reference book to serve both as a guide and an entry point to the field of degenerate mode gyroscopes.

Despite the growing interest in this field, the available information is scattered across a disparate group of conference proceedings and journal papers. For the aspiring scientist/engineer, the scarcity of information forms a large barrier to entry into the field of degenerate mode gyroscopes. This book aims to lower the barrier to entry by providing the reader with a solid understanding of the fundamentals of degenerate mode gyroscopes and its control strategies, as well as providing the necessary know-how and technical jargon needed to interpret future publications in the field.

The book is intended to be a reference material for researchers, scientists, engineers, and college/graduate students who are interested in inertial sensors.

The book may also be of interest to control systems engineers, electrical and electronics engineers, as well as semiconductor engineers who work with inertial sensors. Finally, materials scientists and MEMS production engineers may find the section regarding various fabrication technologies and fabrication defects/energy loss mechanisms interesting.

Doruk Senkal
Andrei M. Shkel

About the Authors

Doruk Senkal

Dr. Senkal has been working on the development of Inertial Navigation Technologies for Augmented and Virtual Reality applications at Facebook since 2018. Before joining Facebook, he was working as a MEMS designer at TDK Invensense, developing MEMS Inertial Sensors for mobile devices.

He received his PhD degree (2015) in Mechanical and Aerospace Engineering from the University of California, Irvine, with a focus on MEMS Coriolis Vibratory Gyroscopes, received his MSc degree (2009) in Mechanical Engineering from Washington State University with a focus on robotics, and received his BSc degree (2007) in Mechanical Engineering from Middle East Technical University.

His research interests, represented in over 20 international conference papers, 9 peer-reviewed journal papers, and 16 patent applications, encompass all aspects of MEMS inertial sensor development, including sensor design, device fabrication, algorithms, and control.

Andrei M. Shkel

Prof. Shkel has been on faculty at the University of California, Irvine, since 2000. From 2009 to 2013, he was on leave from academia serving as a Program Manager in the Microsystems Technology Office of DARPA, where he initiated and managed over $200M investment portfolio in technology development. His research interests are reflected in over 250 publications, 40 patents, and 3 books. Dr. Shkel has been on a number of editorial boards, most recently as Editor of IEEE JMEMS and the founding chair of the IEEE Inertial Sensors (INERTIAL). He has been awarded in 2013 the Office of the Secretary of Defense Medal for Exceptional Public Service, 2020 Innovator of the Year Award, 2009 IEEE Sensors Council Technical Achievement Award, 2008 Researcher of the Year Award, and the 2005 NSF CAREER award. He received his Diploma with excellence (1991) in Mechanics and Mathematics from Moscow State University, PhD degree (1997) in Mechanical Engineering from the University of Wisconsin at Madison, and completed his postdoc (1999) at UC Berkeley. Dr. Shkel is the 2020–2022 President of the IEEE Sensors Council and the IEEE Fellow.

Part I

Fundamentals of Whole-Angle Gyroscopes

1

Introduction

Coriolis Vibratory Gyroscopes (CVGs) are mechanical transducers that detect angular rotation around a particular axis. In its most fundamental form, a CVG consists of two or more mechanically coupled vibratory modes, a forcing system to induce vibratory motion and a sensing system to detect vibratory motion. Angular rotation can be detected by sensing the energy transfer from one vibratory mode to another in the presence of Coriolis forces, Figure 1.1.

Historically, first examples of CVGs can be found in the Aerospace Industry, which were primarily used for navigation and platform stabilization applications. Later, advent of Micro-electromechanical System (MEMS) fabrication techniques brought along orders of magnitude reduction in cost, size, weight, and power (CSWaP), which made CVGs truly ubiquitous. Today CVGs are used in a wide variety of civilian applications, examples include:

- Industrial applications, such as robotics and automation;
- Automobile stabilization, traction control, and roll-over detection;
- Gesture recognition and localization in gaming and mobile devices;
- Optical image stabilization (OIS) of cameras;
- Head tracking in Augmented Reality (AR) and Virtual Reality (VR);
- Autonomous vehicles, such as self-driving cars and Unmanned Aerial Vehicles (UAVs).

1.1 Types of Coriolis Vibratory Gyroscopes

CVGs can be divided into two broad categories based on the gyroscope's mechanical element [1]: degenerate mode (i.e. z-axis) gyroscopes, which have x–y symmetry (Δf of 0 Hz), and nondegenerate mode gyroscopes, which are designed intentionally to be asymmetric in x and y modes ($\Delta f \neq 0$ Hz). Degenerate mode z-axis gyroscopes offer a number of unique advantages compared to nondegenerate vibratory rate gyroscopes, including higher rate sensitivity, ability

Whole-Angle MEMS Gyroscopes: Challenges and Opportunities,
First Edition. Doruk Senkal and Andrei M. Shkel.
© 2020 The Institute of Electrical and Electronics Engineers, Inc. Published 2020 by John Wiley & Sons, Inc.

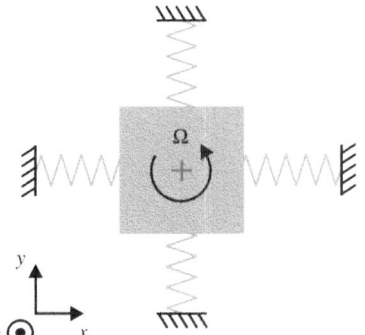

Figure 1.1 Coriolis Vibratory Gyroscopes, in their simplest form, consist of a vibrating element with two or more mechanically coupled vibratory modes. Illustration shows a z-axis gyroscope and its vibratory modes along x- and y-axis.

to implement whole-angle mechanization with mechanically unlimited dynamic range, exceptional scale factor stability, and a potential for self-calibration.

1.1.1 Nondegenerate Mode Gyroscopes

Nondegenerate mode CVGs are currently being used in a variety of commercial applications due to ease of fabrication and lower cost. Most common implementations utilize two to four vibratory modes for sensing angular velocity along one to three axes. This is commonly achieved by forcing a proof mass structure into oscillation in a so-called "drive" mode and sensing the oscillation on one or more "sense" modes. For example, the x-axis of the gyroscope in Figure 1.1, can be instrumented as a drive mode and the y-axis can be instrumented as a sense mode. When a nonzero angular velocity is exerted (i.e. along the z-axis in Figure 1.1), the resultant Coriolis force causes the sense mode (i.e. the mode along the y-axis in Figure 1.1) to oscillate at the drive frequency at an amplitude proportional to input angular velocity.

Resonance frequency of sense modes are typically designed to be several hundreds to a few thousand hertzs away from the drive frequency. The existence of this so-called drive-sense separation (Δf) makes nondegenerate mode gyroscopes robust to fabrication imperfections. However, a trade-off between bandwidth and transducer sensitivity exists since smaller drive-sense separation frequency leads to higher transducer sensitivity, while the mechanical bandwidth of the sensor is typically limited by drive-sense separation (Δf).

Nondegenerate mode gyroscopes are typically operated using open-loop mechanization. In open-loop mechanization, "drive" mode oscillation is sustained via a positive feedback loop. The amplitidue of "drive" mode oscillations are controlled via the so-called Amplitude Gain Control (AGC) loop. No feedback loop is employed on the "sense" mode, which leaves "sense" mode proof mass free to oscillate in response to the angular rate input.

1.1.2 Degenerate Mode Gyroscopes

Degenerate mode gyroscopes utilize two symmetric modes for detecting angular rotation. For an ideal degenerate mode gyroscope, these two modes have identical stiffness and damping; for this reason typically an axisymmetric or x–y symmetric structure is used, such as a ring, disk, wineglass, etc. Degenerate mode gyroscopes are commonly employed in two primary modes of instrumentation: (i) force-to-rebalance (FTR) (rate) mechanization and (ii) whole-angle mechanization.

In FTR mechanization, an external force is applied to the vibratory element that is equal and opposite to the Coriolis force being generated. This is a rate measuring gyroscope implementation, where the magnitude of externally applied force can be used to detect angular velocity. The main benefit of this mode of operation is to boost the mechanical bandwidth of the resonator, which would otherwise be limited by the close to zero drive-sense separation (Δf) of the degenerate mode gyroscope.

In the whole-angle mechanization, the two modes of the gyroscope are allowed to freely oscillate and external forcing is only applied to null the effects of imperfections such as damping and asymmetry. In this mode of operation the mechanical element acts as a "mechanical integrator" of angular velocity, resulting in an angle measuring gyroscope, also known as a Rate Integrating Gyroscope (RIG).

Whole-angle gyroscope architectures can be divided into three main categories based on the geometry of the resonator element: (i) lumped mass systems, (ii) ring/disk systems, and (iii) micro-wineglasses. Ring/disk systems are further divided into three categories: (i) rings, (ii) concentric ring systems, and (iii) disks. Whereas, micro-wineglasses are divided into two categories according to fabrication technology: surface micro-machined and bulk micro-machined wineglass gyroscope architectures, Figure 1.2 [2].

String and bar resonators can also be instrumented to be used as whole-angle gyroscopes, even though these types of mechanical elements are typically not used at micro-scale due to limited transduction capacity. In principle, any axisymmetric elastic member can be instrumented to function as a whole-angle gyroscope.

1.2 Generalized CVG Errors

Gyroscopes are susceptible to a variety of error sources caused by a combination of inherent physical processes as well as external disturbances induced by the environment.

Error sources in a single axis rate gyroscope can be generalized according to the following formula:

$$\tilde{\Omega} = (1 + S_e)\Omega + b_g + n_g, \tag{1.1}$$

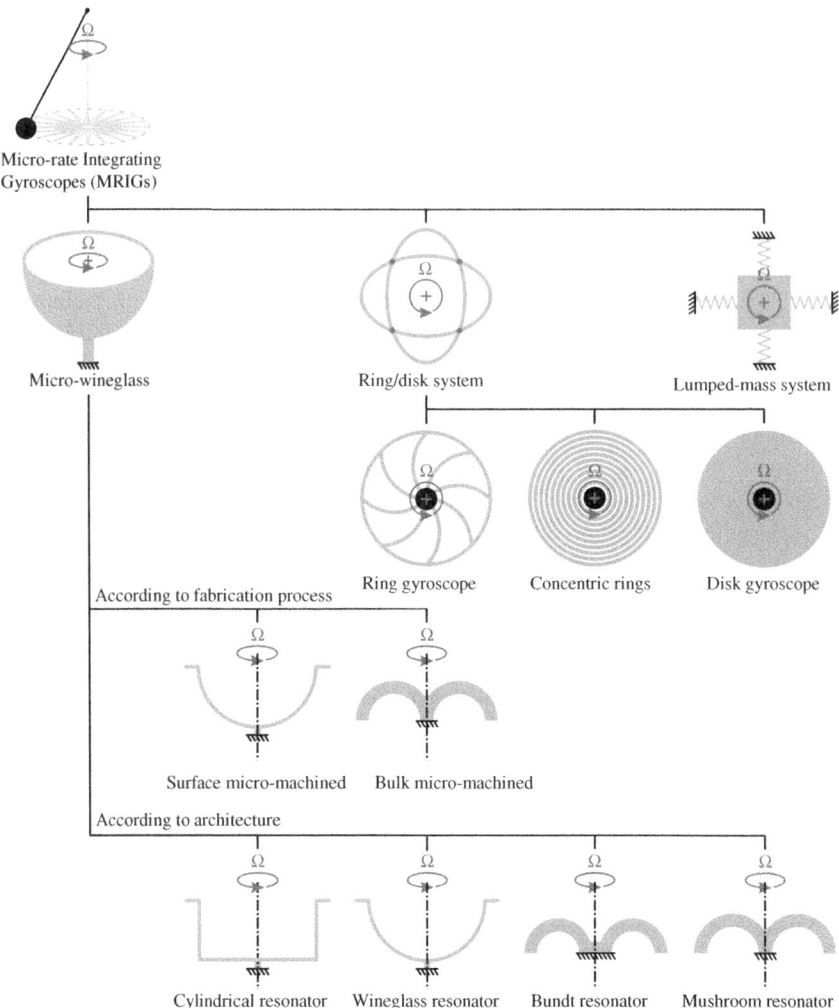

Figure 1.2 Micro-rate integrating gyroscope (MRIG) architectures.

where $\tilde{\Omega}$ is the measured gyroscope output, S_e is scale factor error, b_g is bias error, and n_g is noise. Without loss of generality, for a whole-angle gyroscope the error sources can be written as:

$$\tilde{\theta} = \int ((1 + S_e)\Omega + b_g)dt + n_g, \tag{1.2}$$

where $\tilde{\theta}$ is the measured gyroscope output, corresponding to total angular read-out, including the actual angle of rotation, errors in scale factor, bias, and noise.

1.2.1 Scale Factor Errors

Scale factor (or sensitivity) errors represent a deviation in gyroscope sensitivity from expected values, which results in a nonunity gain between "true" angular rate and "perceived" angular rate. Scale factor errors can be caused by either an error in initial scale factor calibration or a drift in scale factor postcalibration due to a change in environmental conditions, such as a change in temperature or supply voltages, application of external mechanical stresses to the sensing element, or aging effects internal to the sensor, such as a change in cavity pressure of the vacuum packaged sensing element.

1.2.2 Bias Errors

Bias (or offset) errors can be summarized as the deviation of time averaged gyroscope output from zero when there is no angular rate input to the sensor. Aside from initial calibration errors, bias errors can be caused by a change in environment conditions. Examples include a change in temperature, supply voltages or cavity pressure, aging of materials, and application of external mechanical stresses to the sensing element. An additional source of bias errors is external body loads, such as quasi-static acceleration, as well as vibration.

1.2.3 Noise Processes

Noise in gyroscopes can be grouped under white noise, flicker ($1/f$) noise, and quantization noise. The most common numerical tool for representing gyroscope noise processes is Allan Variance.

1.2.3.1 Allan Variance

Originally created to analyze frequency stability of clocks and oscillators, Allan Variance analysis is also widely used to represent various noise processes present in inertial sensors, such as gyroscopes [3]. Allan Variance analysis consists of data acquisition of gyroscope output over a period of time at zero rate input and constant temperature. This is followed by binning the data into groups of different integration times:

$$\overline{\Omega}_n(\tau) = \frac{1}{\tau} \int_{n\tau_0}^{n\tau_0+\tau} \Omega(t) dt, \tag{1.3}$$

where τ_0 is the sampling time, n is the sample number, and $\tau = m\tau_0$ is the bin size. The uncertainty between bins of same integration times is calculated using ensemble average:

$$\sigma^2(\tau) = \frac{1}{2} < (\overline{\Omega}_{n+m}(\tau) - \overline{\Omega}_n(\tau))^2 >. \qquad (1.4)$$

Finally, the calculated uncertainty $\sigma(\tau)$ with respect to integration time (τ) is plotted to reveal information about various noise processes within the gyroscope, Figure 1.3. Sections of the Allan Variance curve and their physical meaning is summarized below [3]:

- **Quantization noise** is due to the conversion of gyroscope output from analog (continuous) signal to digital (countable) signal by Analog-to-Digital Converters (quantization). Quantization noise has a slope of τ^{-1} on the Allan variance graph.
- **Angle Random Walk (ARW)** is caused by white thermomechanical and thermoelectrical noise within the gyroscope, shows up with a slope of $\tau^{-1/2}$. It is usually reported using units $°/\sqrt{h}$ (degrees per square root of hour) or $mdps/\sqrt{Hz}$ (millidegrees per second per square root of hertz).
- **Rate Random Walk (RRW)** is the random drift term within the gyroscope, shows up with a slope of $\tau^{+1/2}$ opposite of ARW.
- **Bias instability** is the lowest point of the Allan variance curve, shows up with a slope of zero. It represents the minimum detectable rate input within the gyroscope and is reported using units $°/h$ (degrees per hour) or mdps (millidegrees per second). Bias instability is limited by a combination of flicker ($1/f$) noise, ARW, and RRW.

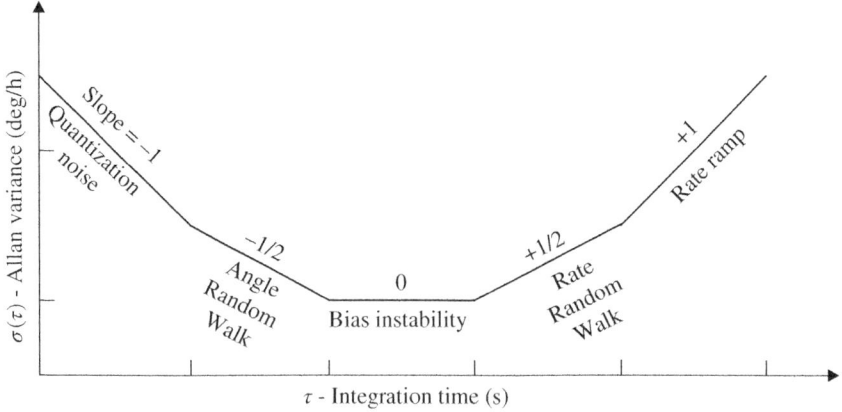

Figure 1.3 Sample Allan variance analysis of gyroscope output, showing error in gyroscope output (deg/h or deg/s) with respect to integration time (s).

- **Rate ramp**, also called the thermal ramp, is caused by temperature changes in the environment, shows up with a slope of τ^{+1}.
- **Periodic oscillations** show up as peaks in the Allan Variance curve with an integration time of 1/frequency (not shown). These oscillations are either caused by a periodic event in sensor electronics or the environment, such as day/night temperature cycles or variations in power supply.

1.3 Overview

The first part of this book focuses on fundamentals of whole-angle gyroscopes, dynamics, mathematical framework, and control strategies. In the second part of the book, conventionally micro-machined 2-D whole-angle gyroscope architectures are reviewed. The final part of the book focuses on 3-D emerging micro-machining technologies for fabrication of whole-angle gyroscope architectures.

In Chapter 2, Foucault Pendulum analogy is introduced as a starting point for whole-angle gyroscope dynamics. Later in the chapter, effects of structural imperfections are introduced, along with requirements for continuous RIG operation.

In Chapter 3, a brief overview of control strategies for CVGs is presented. The remainder of the chapter deals with sustaining oscillation and suppression of error sources in whole-angle gyroscopes, along with methods for tuning and self-calibration.

Chapter 4 provides an overview on two types of conventionally micro-machined 2-D micro-machined whole-angle gyroscope architectures: lumped mass gyroscopes and distributed mass gyroscopes.

In Chapter 5, two examples of 2-D micro-machined whole-angle gyroscope architectures are reviewed: (i) Toroidal Ring Gyroscope (TRG) and (ii) Dual Foucault Pendulum (DFP) gyroscope. The goal of this chapter is to illustrate factors that influence design decisions, fabrication considerations, and characterization methodology.

In Chapter 6, recent advances in 3-D shell micro-technology for fabrication of whole-angle gyroscopes are reviewed. Starting with a brief history of macro-scale shell resonator gyroscopes, the chapter focuses on advances in micro-shell resonator fabrication processes, with an emphasis on transduction mechanisms, characterization techniques, and mechanical properties.

In Chapter 7, the micro-glassblowing paradigm is introduced for wafer-level fabrication of atomically smooth, low internal loss Titania Silicate Glass (TSG) and fused silica 3-D wineglass gyroscopes. Feasibility of the process has been demonstrated by fabrication of fused silica and TSG micro-glassblown structures.

In Chapter 8, various transduction strategies for micro-glassblown wineglass resonators are presented. Two methods of electrostatic transduction are reviewed in this chapter: (i) in-plane electrodes and (ii) out-of-plane electrodes.

Finally, the book is concluded in Chapter 9 with an outlook of future trends.

2

Dynamics

2.1 Introduction to Whole-Angle Gyroscopes

This chapter gives a brief overview of whole-angle gyroscope dynamics and common terms used throughout the book.

2.2 Foucault Pendulum Analogy

Foucault Pendulum, first demonstrated by French physicist Lèon Foucault in 1851, is a device that consists of a proof mass suspended from a long string. The proof mass is free to oscillate in any orientation in the x–y plane. When the device is caused to oscillate at a fixed orientation, the orientation of the oscillation can be observed to slowly precess under the effect of the Coriolis force caused by Earth's rotation. For this reason, the device was used by Foucault to experimentally demonstrate Earth's rotation and is quite possibly the first Rate Integrating Gyroscope (RIG) that was conceived.

If the motion of the Foucault pendulum along the z-axis is sufficiently small, the dynamics can be linearized and approximated by a point mass system in a noninertial frame with equations of motion defined by two coupled spring-mass-dashpot systems, Figure 2.1. This forms a mathematical basis for all Coriolis Vibratory Gyroscopes (CVGs) [4]:

$$\ddot{x} + \frac{2}{\tau_x}\dot{x} + (\omega_x^2 - \eta'\Omega_z^2)x - \eta\dot{\Omega}_z y = \frac{F_x}{m_{eq}} + 2\eta\dot{y}\Omega_z,$$

$$\ddot{y} + \frac{2}{\tau_y}\dot{y} + (\omega_y^2 - \eta'\Omega_z^2)y + \eta\dot{\Omega}_z y = \frac{F_y}{m_{eq}} - 2\eta\dot{x}\Omega_z,$$

(2.1)

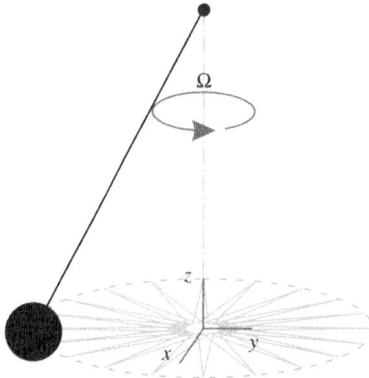

Figure 2.1 Foucault Pendulum is a proof mass suspended from a long string that is free to oscillate in any orientation in the x–y plane. Once started, the straight line oscillation pattern precesses under the effect of the Coriolis force, caused by Earth's rotation.

where τ_x and τ_y are the highest and the lowest energy decay time constants along the x and y axes, respectively. Energy decay time constant is defined as the time that takes for vibration amplitude of the resonator to decay down to $1/e \approx 36.8\%$ of initial amplitude.

The parameters ω_x and ω_y are the highest and lowest resonance frequencies, in rad/s, along the x and y axes. Throughout the book, f_x and f_y are also used for the resonance frequencies; in this case the units would be in Hz or kHz ($f = 2\pi\omega$).

Finally, m_{eq} is the equivalent mass of the gyroscope, which is the mass that is sensitive to the Coriolis force. F_x and F_y are the forcing functions along the x and y axes, respectively. Ω_z is the angular velocity along the z axis and η is the angular gain factor of the gyroscope. For an ideal gyroscope $\eta = 1$, but typically $\eta < 1$, and this is due to the additional mass in the system that does not contribute to the Coriolis force. It must also be noted that due to slow angular rotation rates compared to gyroscope resonance frequency ($\omega \gg \Omega$) and small amplitude of oscillation, the contribution of higher order inertial terms, proportional to Ω^2, are generally negligible. In our discussion, we also assume that the input angular acceleration $\ddot{\Omega}$ is zero, that is, the object is rotating with constant angular velocity. Unlike the inertial forces that are proportional to Ω^2 and $\ddot{\Omega}$, the Coriolis force has a unique skew-symmetric topology and is equal to the mass multiplied to the cross product of the angular velocity vector Ω and the linear velocity vector, which is measured relative to the noninertial coordinate frame.

2.2.1 Damping and Q-factor

Energy decay time constant ($\Delta\tau^{-1}$) was introduced in the previous section. Another closely related quantity to the energy decay time constant is the quality factor (Q-factor), which is introduced in this section.

The Q-factor is a dimensionless quantity that represents the rate of energy loss of a system per cycle with respect to the stored energy of the system:

$$Q = \frac{\text{Stored energy}}{\text{Energy dissipated per cycle}}. \quad (2.2)$$

In other words, it is a measure of energy dissipation to stored energy. Being a dimensionless quantity, the Q-factor can be used to represent the energy loss in a wide variety of physical domains, including mechanical, optical, electrical, electro-mechanical, etc. However, in this book it is exclusively used to define the internal energy loss of the mechanical resonator within the system.

For systems with sufficiently low energy dissipation, it can be tied to the time constant of the resonator (τ) and resonance frequency (f) through:

$$Q = \frac{\tau}{\pi f}. \quad (2.3)$$

Finally, it can be extracted from frequency response of a second-order system using the center frequency and full width of the resonator at −3 dB point:

$$Q = \frac{\text{Center frequency}}{\text{Full width at} - 3 \text{ dB point}}. \quad (2.4)$$

Maximization of the Q-factor is key to enhancing performance of vibratory MEMS devices in demanding signal processing, timing, and inertial applications [5]. In this section, various dissipation mechanisms contributing to the overall Q-factor are reviewed.

The total Q-factor of a vibratory structure can be calculated from the contribution of individual dissipation mechanisms in a manner analogous to solving a parallel resistor network, Eq. (2.5). For this reason, the total Q-factor is dominated by the dissipation mechanism with the lowest Q-factor (weakest link). Some known and frequently considered mechanisms that contribute to the total quality factor are:

$$Q_{\text{total}}^{-1} = Q_{\text{visc}}^{-1} + Q_{\text{anchor}}^{-1} + Q_{\text{mat}}^{-1} + Q_{\text{surf}}^{-1} + Q_{\text{coupling}}^{-1} + Q_{\text{etc}}^{-1}. \quad (2.5)$$

In order to optimize the Q-factor, all loss mechanisms affecting the system need to be individually addressed. These are viscous damping (Q_{visc}), anchor losses (Q_{anchor}), material losses (Q_{mat}), surface losses (Q_{surf}), mode coupling losses (Q_{coupling}), and additional dissipation mechanisms (Q_{etc}).

2.2.1.1 Viscous Damping

Viscous damping (Q_{visc}) is typically the most dominant dissipation mechanism at atmospheric conditions, limiting the maximum Q-factor to a few hundreds in air. In MEMS gyroscopes, analysis of viscous damping is complicated due to: (i) relatively complex device geometry, and (ii) a break-down in continuum approximation of fluid mechanics for the characteristic dimensions considered in MEMS. As the mean free path of the gas molecules approaches the device characteristic length, the gas needs to be treated as discrete particles (molecular flow regime), as opposed to continuous mass (viscous flow regime). Although a complete analysis of viscous damping in MEMS gyroscopes is outside the scope of this text, means for identifying the correct flow regime is described below to help interested readers in choosing the appropriate analytical framework.

The ratio of mean free path to critical dimension, also known as Knudsen number, can be used to identify the flow regime of the device [6]:

$$K_n = \frac{\lambda}{d_c}, \tag{2.6}$$

where λ is the mean free path of the gas molecules and d_c is the characteristic length of the mechanical element, typically the capacitive gaps of the gyroscope. The mean free path λ can be calculated using [6]:

$$\lambda = \frac{RT}{\sqrt{2}\pi d_g^2 NP}, \tag{2.7}$$

where R is the universal gas constant, T is the temperature measured in kelvin, d_g is the effective gas molecule diameter, N is the Avogadro number, and P is the pressure within the device cavity measured in pascal. Once the Knudsen number of the device is known, the correct flow regime can be identified using:

- $K_n > 10$: molecular flow regime, where gas needs to be treated as discrete particles;
- $0.1 < K_n < 10$: transition region, where continuum flow approximation starts breaking down;
- $K_n < 0.1$: continuum flow regime, where gas needs to be treated as a continuous mass.

For most gyroscopes, viscous damping can be reduced to a negligible level by packaging the gyroscope in vacuum and depositing getter materials inside the package to capture excess gas molecules. Alternatively, "getter-less" packaging techniques, such as Epitaxial Encapsulation (Episeal) process, described in Chapter 4, can be used to achieve pressure levels on the order of 1 Pa without the need for getter materials [7]. In most CVGs, the design objective is to reduce air damping to negligible levels, and this is feasible for many modern vacuum packaging techniques [8].

2.2.1.2 Anchor Losses

Anchor losses (Q_{anchor}) are caused by acoustic energy loss into the substrate. Anchor losses can be minimized by decoupling the resonator motion from the substrate by (i) achieving linear and angular momentum balance in the primary modes of the gyroscope and (ii) anchoring the gyroscope at the nodal points (zero displacement) of the primary modes. Although there are no closed-form solutions for anchor loss in complex geometries, Perfectly Matched Layers (PMLs) can be used in Finite Element Analysis (FEA) to calculate anchor losses. PMLs simulate anchor losses by creating an "absorption layer" around the gyroscope, preventing outbound acoustic energy from reflecting back into the gyroscope. In order to correctly estimate anchor losses, care must be taken in the implementation of PMLs:

- A PML with minimum size of at least half acoustic wavelength needs to be used for absorption. For gyroscopes with primary modes in the kHz range, this can create PML layers with several meters in size, greatly increasing computational complexity;
- PML geometry needs to be chosen so that acoustic waves in all directions can be absorbed. For this reason, a hemisphere for 3-D simulations and a semicircle for 2-D simulations is a common choice for PML geometry;
- PML needs to be meshed with a sufficiently dense mesh to fully absorb acoustic waves propagating into the substrate. Correct mesh density can be evaluated via a "mesh convergence" study, in which the mesh density is varied and the anchor loss magnitude is calculated for each unique mesh density. "Mesh convergence" is achieved when further increase in mesh density creates a negligible change in anchor loss magnitude.

2.2.1.3 Material Losses

Material losses (Q_{mat}) can be divided into several individual loss mechanisms. Thermoelastic Dissipation (TED) is a source of internal loss present in all materials, caused by the interaction of temperature fluctuations and oscillations within a vibratory structure [5, 9]. Structures that are vibrating exhibit stress/strain gradients, which in turn create localized temperature gradients within the structure. These temperature gradients inevitably result in heat transfer within the structure and an irreversible conversion of vibratory energy into heat. The degree of coupling between the temperature gradients and stress/strain gradients is controlled by the material's Coefficient of Thermal Expansion (CTE), which is an intrinsic property of the material [10, 11].

TED within a vibratory structure depends on a wide variety of factors, such as resonator geometry, vibration frequency, and resonator temperature. In order to reduce TED: (i) resonant frequency of the gyroscope can be chosen such that the frequency difference between the thermal and mechanical modes is maximized and (ii) low CTE materials, such as fused silica, can be used for fabrication of the mechanical element. Low CTE materials can provide orders of magnitude reduction in TED due to a reduction in coupling between thermal and mechanical domains.

Additional material losses are caused by microscopic effects, such as presence of foreign materials within the matrix of the resonator material and lattice defects at grain boundaries [12]. These effects can be minimized by using a high purity material such as fused silica or silicon.

2.2.1.4 Surface Losses
Surface losses (Q_{surf}) are mainly caused by energy dissipation due to surface defects and metallization losses [12]. The effect is especially pronounced in micro-scale devices due to a significantly larger surface-to-volume ratio.

Surface losses can be minimized by (i) making sure the resonator surfaces are smooth and free of contaminants, and (ii) by keeping the thickness of the metal layer small with respect to the resonator thickness [13].

2.2.1.5 Mode Coupling Losses
Mode coupling losses ($Q_{coupling}$) occur when one or more low Q-factor modes are close to the primary resonance modes of the gyroscope [6]. Combined quality factor of multiple factors can be calculated using:

$$\frac{1}{Q_{coupling}} = \frac{\alpha_1}{Q_1} + \frac{\alpha_2}{Q_2}, \tag{2.8}$$

where Q_1 and Q_2 are Q-factors of each mode and α_1 and α_2 are the ratio of stored energy in each mode. In order to minimize mode coupling losses, mechanical element of the gyroscope should be designed such that the resonance frequency of the low Q-factor "parasitic" modes is not close in frequency to the primary modes of the gyroscope.

2.2.1.6 Additional Dissipation Mechanisms
Additional dissipation mechanisms (Q_{etc}), such as Akheiser dissipation, have typically very high Q-factors at kHz range and are not taken into account [14].

2.2.2 Principal Axes of Elasticity and Damping
Due to fabrication imperfections, x and y energy decay time constants and resonance frequencies along the x and y axes differ from each other. The so-called aniso-damping (damping split) and aniso-elasticity (frequency split) can be represented as:

$$\Delta\omega = \omega_x - \omega_y \text{ or } \Delta f = f_x - f_y, \tag{2.9}$$

$$\Delta\tau^{-1} = \tau_x^{-1} - \tau_y^{-1}, \tag{2.10}$$

where the center frequency and nominal energy decay constant are:

$$\omega = \frac{\omega_x + \omega_y}{2} \text{ or } f = \frac{f_x + f_y}{2}, \tag{2.11}$$

$$\tau = \frac{\tau_x + \tau_y}{2}. \tag{2.12}$$

The axes where the highest and lowest resonance frequencies and energy decay time constant occur are called the principal axes of elasticity and damping, respectively. So far, it was assumed that the principal axes of elasticity and

2.2 Foucault Pendulum Analogy

(a) Principal axes of elasticity

(b) Principal axes of damping

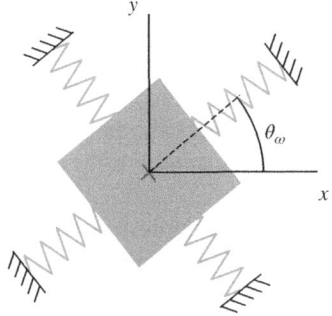
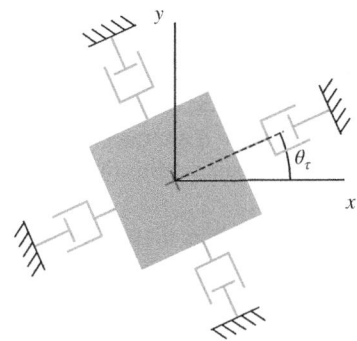

Figure 2.2 Misalignment in principal axes of (a) elasticity and (b) damping of the gyroscope with respect to the forcer/pick-off axis. For an actual gyroscope, the principal axes of elasticity and damping are not necessarily aligned ($\theta_\omega \neq \theta_\tau$).

damping are aligned with the x–y coordinate axes, Eq. (2.1). This is never the case for an actual gyroscope. Due to fabrication imperfections, as well as electrode misalignments, the principal axes of elasticity and damping are generally misaligned to the forcer/pick-off axis. Effect of such a misalignment is shown in Figure 2.2 for principal axis of elasticity and damping.

The principal axes of elasticity are mutually orthogonal, and the same can be said about the principal axes of damping. This effect can be represented using an orthogonal transformation by introducing the rotation angles θ_τ and θ_ω [4]. Here, θ_τ represents the orientation of the τ_x axis and θ_ω represents the orientation of the ω_y axis. For an actual gyroscope, the principal axes of elasticity and damping are not necessarily aligned ($\theta_\omega \neq \theta_\tau$).

After the introduction of aniso-elasticity, aniso-damping terms, and their orientation with respect to forcer/pick-off axis, the equations of motion become [4]:

$$\ddot{x} + \frac{2}{\tau}\dot{x} + \Delta\tau^{-1}(\dot{x}\cos(2\theta_\tau) + \dot{y}\sin(2\theta_\tau)) + (\omega_x^2 - \eta'\Omega_z^2)x - \eta\dot{\Omega}_z y$$
$$-\omega\Delta\omega(x\cos(2\theta_\omega) + y\sin(2\theta_\omega)) = \frac{F_x}{m_{eq}} + 2\eta\dot{y}\Omega_z,$$

$$\ddot{y} + \frac{2}{\tau}\dot{y} + \Delta\tau^{-1}(\dot{x}\sin(2\theta_\tau) - \dot{y}\cos(2\theta_\tau)) + (\omega_y^2 - \eta'\Omega_z^2)y + \eta\dot{\Omega}_z x$$
$$-\omega\Delta\omega(x\sin(2\theta_\omega) - y\cos(2\theta_\omega)) = \frac{F_y}{m_{eq}} - 2\eta\dot{x}\Omega_z.$$
(2.13)

It is important to note that after the introduction of the misalignment errors in principal axis of elasticity and damping, the two vibratory modes of the gyroscope are not only coupled by the Coriolis forces, but also through the aniso-elasticity and aniso-damping terms $\Delta\omega$ and $\Delta\tau^{-1}$, respectively.

2.3 Canonical Variables

For nonzero initial velocity and in the absence of rotation, damping, and aniso-elasticity, the equivalent point-mass system described in Eq. (2.13) would trace a constant elliptical orbit, as shown in Figure 2.3 [15].

In Figure 2.3, a and q are major and minor radii of the elliptical orbit, θ is the orientation of the major radius (also known as pattern angle), and φ is the phase of the oscillation.

For an ideal RIG, where there is no damping or aniso-elasticity, the major radius of the elliptical trajectory a is expected to remain constant and minor radius q is expected to be constant and zero, while the orientation of major radius (pattern angle) precesses in time under the effect of the Coriolis force, according to:

$$\theta = -\eta \int \Omega dt = -\eta \theta_{\text{actual}}, \tag{2.14}$$

where η is the angular gain factor defined by the mass distribution of the gyroscope and Ω is the angular velocity measured by the gyroscope. As can be seen in Eq. (2.14), the orientation of the precession pattern θ is directly related to the integral of the angular velocity of the gyroscope through angular gain factor η. Due to this close relationship, the orientation of precession pattern (θ) becomes a measure of actual rotation measured by the gyroscope (θ_{actual}). Hence, this mode of mechanization is called RIG or whole-angle gyroscope operation.

2.4 Effect of Structural Imperfections

So far, the canonical variables were described for an ideal gyroscope, where the damping and aniso-elasticity are zero. However, due to imperfections within the system, such as a nonzero dissipation term as well as aniso-elasticity and

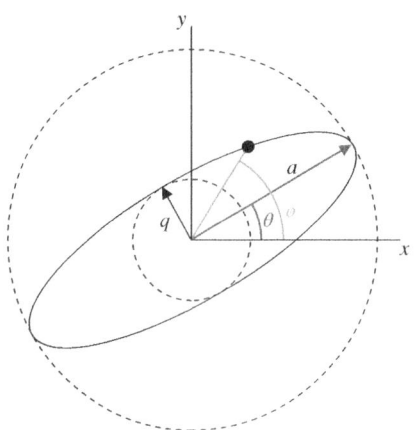

Figure 2.3 Elliptical orbit of a CVG and the canonical variables.

2.4 Effect of Structural Imperfections

aniso-damping, these variables a, q, and φ vary slowly in time. In addition to this, the pattern angle θ is no longer only a function of the Coriolis input (Ω), but is affected by aniso-elasticity and aniso-damping.

Despite this slow variation, the terms a, q, φ, and θ can be used to describe the state of the CVG, particularly if the aniso-damping and aniso-elasticity terms are sufficiently small with respect to ω, such that $\Delta\omega \ll \omega$ and $\Delta\tau^{-1} \ll \omega$. In this case, the state of the gyroscope can be defined by using the four canonical variables: pattern angle θ, the phase angle φ, as well as the energy term [4]:

$$E = a^2 + q^2, \tag{2.15}$$

and the quadrature term:

$$Q = 2aq. \tag{2.16}$$

The so-called quadrature error (Q) is the unwanted component of the oscillation, which manifests as a result of aniso-elasticity within the mechanical element. The quadrature error (Q) can inadvertently interfere with estimation of the pattern angle, due to imperfections in pick-off and control electronics. Moreover, if the quadrature term is allowed to grow to a point, such that the elliptical orbit of the gyroscope degenerates into a circle ($a = q$), it becomes impossible to estimate the pattern angle θ, Figure 2.3.

As described in the previous section, for an ideal gyroscope, the time evolution of these canonical variables would be [4]:

$$\begin{aligned}&\dot{E} = 0,\ E \neq 0, \\ &\dot{Q} = 0,\ Q = 0, \\ &\dot{\theta} = -\eta\Omega.\end{aligned} \tag{2.17}$$

However, when the structural imperfections are introduced, the time evolution of the canonical variables become [4]:

$$\begin{aligned}\dot{E} = &-\frac{2}{\tau}E - \Delta\tau^{-1}\cos 2(\theta - \theta_\tau)\sqrt{E^2 - Q^2} \\ &-\frac{a}{\omega}f_{as}\cos\delta\phi + \frac{q}{\omega}(f_{qc}\cos\delta\phi + f_{qs}\sin\delta\phi), \\ \dot{Q} = &-\frac{2}{\tau}Q - \Delta\omega\sin 2(\theta - \theta_\omega)\sqrt{E^2 - Q^2} \\ &+\frac{a}{\omega}(f_{qc}\cos\delta\phi + f_{qs}\sin\delta\phi) - \frac{q}{\omega}f_{as}\cos(\delta\phi), \\ \dot{\theta} = &-\eta\Omega + \frac{1}{2}\Delta\tau^{-1}\sin 2(\theta - \theta_\tau)\frac{E}{\sqrt{E^2 - Q^2}} \\ &+\frac{1}{2}\Delta\omega\cos 2(\theta - \theta_\omega)\frac{Q}{\sqrt{E^2 - Q^2}} \\ &-\frac{a}{2\omega\sqrt{E^2 - Q^2}}(f_{qs}\cos\delta\phi - f_{qc}\sin\delta\phi) + \frac{q}{2\omega\sqrt{E^2 - Q^2}}f_{as}\sin\delta\phi,\end{aligned} \tag{2.18}$$

where f_{as}, f_{ac}, f_{qs} and f_{qc}, are external forcing functions that can be potentially used to control the shape and orientation of the precession pattern of the gyroscope and null the effects of structural imperfections, such as aniso-damping and aniso-elasticity. Here the subscripts a and q denote forces acting on major and minor radii of the ellipsoid, respectively. Whereas the subscripts c and s denote forcing functions that are in phase and in quadrature to the oscillation of the mechanical element, respectively. The forcing functions, which are necessary to achieve continuous RIG operation are discussed in detail in Chapter 3. However, it is important to note here that the forcing functions f_{as}, f_{ac}, f_{qs}, and f_{qc} are used to control the gyroscope, acting on multiple canonical variables simultaneously, Eq. (2.18). This makes it extremely important that the cross-talk between the four control loops, as well as errors in pattern angle estimation (θ), is minimized in order to eliminate unwanted perturbations to the precession pattern.

As can be seen in Eq. (2.18), in order to achieve the ideal behavior described in Eq. (2.17), where the amplitude is constant, the quadrature error is zero and the pattern angle is only a function of Coriolis forces, we would need to achieve:

$$\tau = \infty, \tag{2.19}$$

$$\Delta\omega = 0, \tag{2.20}$$

$$\Delta\tau^{-1} = 0. \tag{2.21}$$

This brings us to the following important criteria regarding the mechanical element of an RIG:

1) In order to minimize the unwanted drift in the angle of precession pattern (θ), the structural imperfections $\Delta\omega$ and $\Delta\tau^{-1}$ need to be minimized.
2) To keep the quadrature error small, the aniso-elasticity ($\Delta\omega$) needs to be minimized.
3) Finally, the energy decay time constant τ need to be maximized, so that the energy pumping necessary to sustain oscillation amplitude (E) can be minimized.

2.5 Challenges of Whole-Angle Gyroscopes

Despite potential advantages of whole-angle gyroscopes, achieving the required degree of stiffness and damping symmetry through micro-machining processes remains to be a challenge. For example, RIG performance relies heavily on the stiffness and damping symmetry (Δf and $\Delta\tau$) between the two degenerate modes [16]. For macro-scale RIGs, this frequency symmetry is obtained through a combination of precision machining processes (with 10^{-6} relative tolerance) and postfabrication trimming of resonators [17]. Whereas conventional micro-machining processes are generally associated with moderate fabrication accuracy ($10^{-2} - 10^{-4}$ relative tolerance) and structures that are typically not suitable for postfabrication polishing and trimming. Factors such as mold nonuniformity, alignment errors or high surface roughness, and granularity of deposited thin films have so far prevented fabrication of high precision 3-D micro-shell structures using MEMS techniques.

In addition to high precision fabrication processes, maximization of the quality factor is key to enhancing performance of vibratory MEMS devices in demanding signal processing, timing, and inertial applications [5]. For example, timing devices rely heavily on high Q-factors for low close-to-carrier phase noise and low energy consumption [18]. Whereas, devices such as high performance RIGs and mode-matched rate gyroscopes rely heavily on long energy decay times and high Q-factors [16]. Current MEMS fabrication techniques limit the maximum achievable Q-factor by restricting the material choice to a few materials. Available materials, such as single-crystal silicon, have relatively high CTE (~3 ppm/°C [19]) and consequently, high TED [9]. Materials with low CTE, such as fused silica (0.5 ppm/°C), ultra low expansion titania silicate glass (ULE TSG, 0.03 ppm/°C) and polycrystalline diamond (1.2 ppm/°C) can provide a dramatic increase in fundamental Q_{TED} limit. However, when compared to silicon, TSG and fused silica dry etching suffers from an order of magnitude higher surface roughness, lower mask selectivity (~1 : 1 for photoresist), and lower aspect ratio, <5 : 1 [20, 21].

3

Control Strategies

In this chapter, a brief introduction to rate gyroscope control strategies is reviewed first. This is followed by control strategies that are specific to the whole-angle gyroscope implementations. Finally, experimental results on parametric drive of whole-angle gyroscopes and parameter identification are discussed.

3.1 Quadrature and Coriolis Duality

Quadrature errors were introduced in Chapter 2 as a part of the whole-angle gyroscope dynamics. The quadrature and Coriolis duality is a fundamental part of all Coriolis Vibratory Gyroscopes (CVGs) regardless of mechanization type. The so-called quadrature error manifests due to stiffness imperfections of the electro-mechanical element and is the result of drive displacement coupling into the sense axis. Whereas, the so-called Coriolis signal, the quantity to be measured, is generated by modulation of the drive velocity by Coriolis forces. Since the quadrature signal is generated via drive displacement and the Coriolis signal is generated via drive velocity, both signals occur at drive frequency with a 90° phase delay between them, Figure 3.1. This 90° phase separation makes it possible to isolate the desired Coriolis signal from the unwanted quadrature signal via synchronous demodulation.

Synchronous demodulation is achieved by mixing the Coriolis and quadrature signals with an unmodulated harmonic signal at drive frequency. By aligning the phase of the unmodulated signal with Coriolis or quadrature, the amplitude of the respective signal can be estimated.

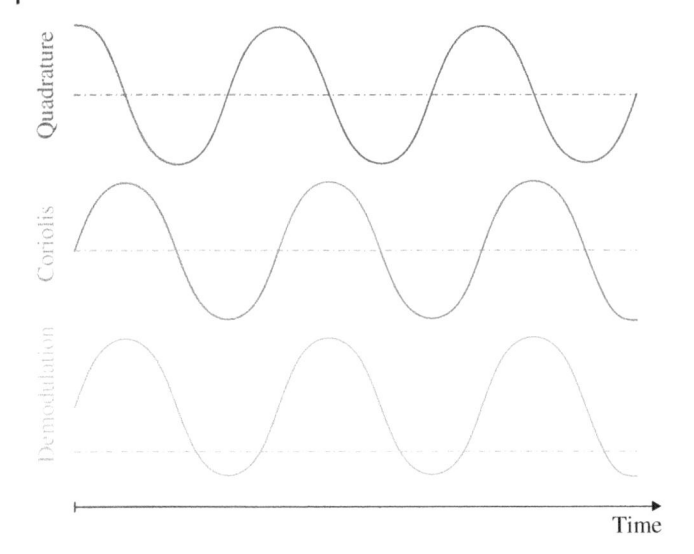

Figure 3.1 Quadrature signal can be separated from Coriolis signal via synchronous demodulation.

3.2 Rate Gyroscope Mechanization

Rate gyroscopes can be instrumented using two distinct mechanization schemes: open-loop and force-to-rebalance (FTR). Both of these mechanization strategies are in use today and the correct choice of CVG mechanization depends largely on the application.

3.2.1 Open-loop Mechanization

Open-loop mechanization is the most common mechanization strategy in use today. Due to the ease of implementation and a relatively lower reliance on initial calibration, it is widely used for gyroscopes intended for consumer and automotive applications, which utilize nondegenerate mode gyroscopes with a nonzero frequency split between drive and sense modes.

The control system consists of four fundamental subsystems: (i) drive mode oscillator, (ii) Amplitude Gain Control (AGC), (iii) Phase Locked Loop (PLL)/demodulation, and (iv) quadrature cancellation, Figure 3.2.

3.2.1.1 Drive Mode Oscillator
The drive mode oscillator circuit is responsible for sustaining oscillation by providing positive feedback to the drive mode of the gyroscope. This circuit is responsible for detecting drive mode oscillation and applying a phase shifted excitation force

3.2 Rate Gyroscope Mechanization | 25

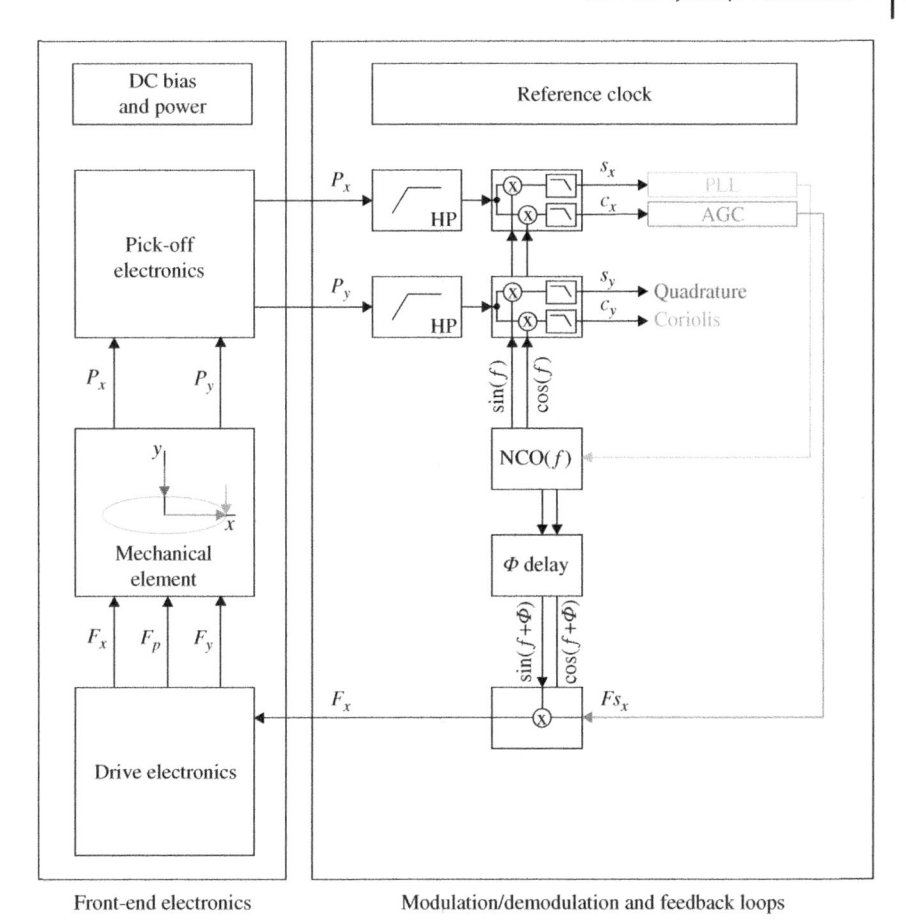

Figure 3.2 Open-loop mechanization utilizes no feedback loop in the sense mode (y).

to the drive mode, such that the total phase delay within the loop is 360°. Analog implementations consist of a front-end amplifier for detection, a phase shifter, and drive force electronics. Digital implementations also exist, which move the phase shifting operation into a digital circuit by adding an Analog to Digital Converter (ADC) for quantization of drive motion and a Digital to Analog Converter (DAC) for excitation.

Initial factory calibration of this subsystem commonly consists of gain adjustment of the front-end amplifier and adjustment of the phase delay within the phase shifter. Failing to calibrate these parameters may result in incorrect drive amplitude or out-of-phase drive excitation.

3.2.1.2 Amplitude Gain Control

Amplitude Gain Control (AGC) loop works in tandem with the drive mode oscillator to keep the drive oscillation amplitude stable. Since gyroscope's scale factor is dependent on drive velocity in open-loop mechanization, having a stable drive amplitude is required to have a good scale factor stability. This loop consists of an amplitude detection subsystem for sensing drive oscillation magnitude and a Proportional Integral Derivative (PID) controller to control the excitation force exerted by the drive mode oscillator. Factory calibration of this subsystem is achieved by adjusting the target drive excitation level and tuning of the PID parameters.

3.2.1.3 Phase Locked Loop/Demodulation

Phase Locked Loop (PLL) is responsible for tracking drive oscillation with the ultimate goal of demodulating the sense mode output at the drive frequency in order to recover rate information from the sense mode output.

Sense mode output contains two AC signals with strictly 90° phase delay in between the Coriolis signal and the quadrature signal. The quantity to be measured is referred as the Coriolis signal, amplitude of which changes as a result of the rate input to the gyroscope's sensitive axis. The unwanted signal is referred to as the quadrature signal and is born of imperfections within the CVG mechanical element. The quadrature signal can often be orders of magnitude larger than the Coriolis signal and its amplitude changes as a result of environmental factors such as temperature and package stress. A demodulation scheme is used to detect the amplitude of the Coriolis signal while rejecting a bulk of the quadrature signal. For this reason, the selection of correct demodulation phase is very important, since an incorrectly selected demodulation phase may cause an unwanted quadrature signal to have a large influence on gyroscope's output, degrading the in-run bias stability.

3.2.1.4 Quadrature Cancellation

Depending on the severity of fabrication imperfections, the quadrature signal can be several orders of magnitude larger than the Coriolis signal. This poses a challenge for the front-end-amplifier, since a large unwanted signal in the sense mode output can degrade signal-to-noise ratio or even cause saturation of the front-end amplifier. For this reason, a quadrature cancellation subsystem is utilized in the open-loop gyroscopes. The most common implementation consists of a feed-through capacitor placed between the drive excitation and the sense mode output. Using the correct capacitance value, the quadrature signal can be canceled to the first order, with only a small residual quadrature remaining at the amplifier input. Typically, a capacitor array can be used for this purpose; by switching different sized capacitors in and out, the equivalent capacitance value can be adjusted to the correct value during factory calibration.

3.2.2 Force-to-rebalance Mechanization

Force-to-rebalance (FTR) mechanization takes the open-loop mechanization one step further by including closed-loop control of the sense mode, while retaining drive mode oscillator, AGC, PLL, and demodulation subsystems, Figure 3.3. Most commercial mode-matched gyroscopes operate using FTR mechanization due to the benefits of increased sensor bandwidth, improved linearity, and a better scale factor stability. Gyroscopes that use FTR mechanization are also referred to as "closed loop gyroscopes" or "force-feedback gyroscopes."

The goal of FTR mechanization is to actively force the sense mode oscillation to zero by utilizing a separate excitation channel for the sense mode. Since during normal operation the sense mode output amplitude will be zero, the excitation force required to drive the sense mode to zero becomes a proxy for the sensing rate input.

3.2.2.1 Force-to-rebalance Loop

To be able to implement the FTR mechanization, the capability to excite the sense mode is required. This can be achieved by adding a sense mode excitation scheme, such as the sense mode forcer electrodes. The FTR loop is created by introducing a PID controller, which can control the amplitude of the forcing signal based on the Coriolis signal amplitude. The demodulated Coriolis signal is fed to the PID controller as the quantity to be controlled and output of the PID controller becomes the forcing signal magnitude which can drive sense mode to zero displacement. Once the required forcing signal magnitude is calculated by the PID controller, it goes through a remodulation step in order to convert to an AC signal at drive frequency and in phase with the actual Coriolis force, Figure 3.3.

There are several advantages to FTR mechanization over open-loop mechanization:

1) **Increased sensor bandwidth:** For open-loop mechanization the bandwidth is limited by $f_{BW} \leq \Delta f$ for $1/\tau_y \ll f_{BW}$ [22]. Mode-matched gyroscope ($\Delta f \approx 0$) with high sense Q-factor demonstrate close to zero bandwidth in the open-loop mechanization. FTR mechanization bypasses this constraint by actively forcing the sense mode oscillation amplitude to zero, increasing the sensor bandwidth to the bandwidth of the FTR loop. This is perhaps the most important advantage of the FTR mechanization, as it enables the use of mode-matched devices in real-life applications, where a nonzero sensor bandwidth is needed.
2) **Improved sensor linearity:** For high angular rate inputs to a mode-matched gyroscope in the open-loop mechanization, sense mode displacement (and

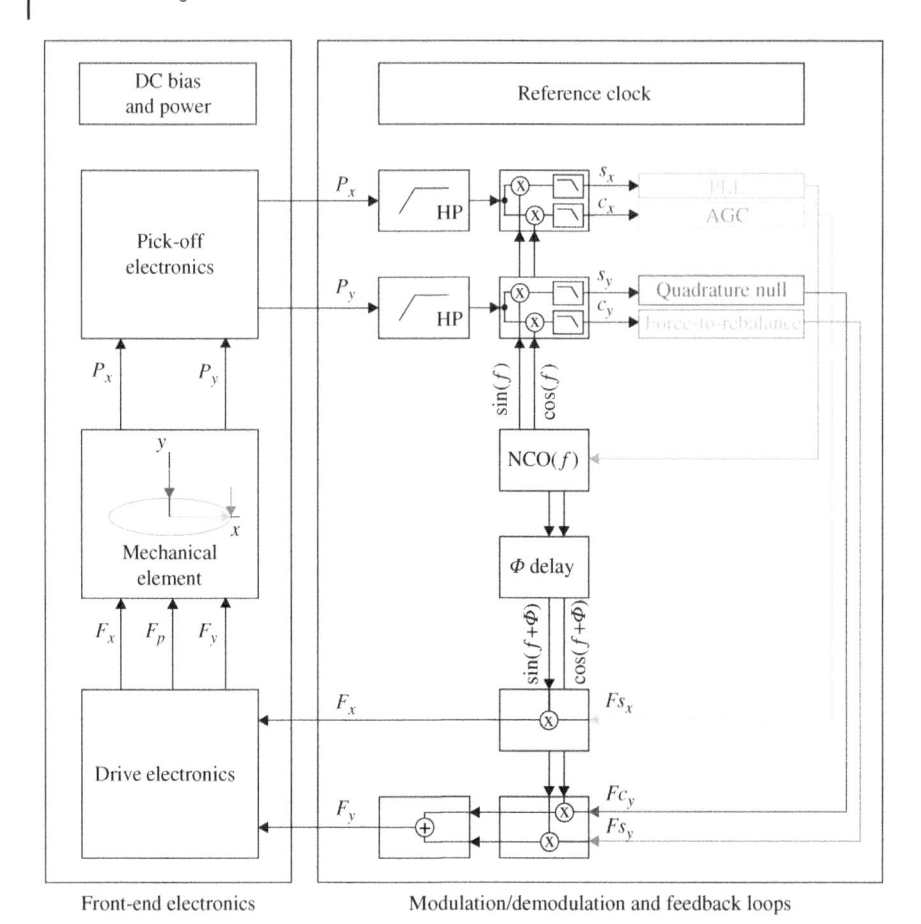

Figure 3.3 Force-to-rebalance utilizes feedback loops in both the drive (x) and the sense modes (y).

velocity) can reach an amplitude comparable to drive displacement. Large sense mode displacement effectively changes the angle of the drive velocity vector with respect to the sense mode and as a result the Coriolis force direction starts diverging from the sense mode direction, resulting in a reduction of sensitivity for high angular rate inputs. FTR mechanization eliminates this problem by actively forcing the sense mode to zero displacement.

3) **Better scale factor stability:** In the open-loop mechanization, sense mode Q-factor contributes to the gyroscope's scale factor. This effect is especially pronounced for mode-matched gyroscopes, which have a one-to-one relationship between sense mode Q-factor and the gyroscope scale factor. This is an undesirable trait in gyroscopes, since the sense mode Q-factor can be affected by

environmental factors, such as temperature and aging, resulting in the scale factor drift. FTR mechanization bypasses this problem by using the "force required to null sense mode displacement" as a proxy for rate input, as opposed to using the sense mode displacement itself, which is affected by the sense mode gain.

Compared to the open-loop mechanization, the main disadvantages of the FTR mechanization is increased complexity, cost, and power consumption.

The FTR loop is also used in the whole-angle mechanization for the purpose of pattern angle control. Although the functionality is similar, the implementation for the whole-angle mechanization is significantly different and will be covered in the next section.

3.2.2.2 Quadrature Null Loop

The quadrature null loop is responsible for actively forcing the quadrature displacement within the sense mode to zero. Unlike the open-loop quadrature cancellation, which aims to cancel only the electrical signal generated due to quadrature, the quadrature null loop physically forces the sense mode quadrature displacement to zero. For this reason, quadrature null loop has the additional advantage of reacting to changes in the quadrature magnitude in a closed-loop fashion, making corrections in real time and potentially improving the gyroscope stability.

Implementation of the quadrature null loop is very similar to the FTR loop, in the sense that a modulation step is used to force the sense mode at the drive frequency and an associated PID loop to adjust the amplitude of the applied force based on changes in the quadrature signal. The main difference between the quadrature null loop and the FTR loop is a 90° phase difference in demodulation and remodulation steps. As a result of this 90° phase difference the quadrature signal to be fed to the PID controller and the remodulation is done such that the forcing signal is in phase with the quadrature force phase.

Similar to the FTR loop, the quadrature null loop is also used as a part of the whole-angle mechanization.

3.3 Whole-Angle Mechanization

As described in Chapter 1, in the whole-angle mechanization, the mechanical element acts as a "mechanical integrator" of angular velocity, resulting in an angle measuring gyroscope, also known as a Rate Integrating Gyroscope (RIG). To achieve this behavior, the two modes of the gyroscope are allowed to freely oscillate and external forcing is only applied to null the effects of imperfections, such as damping and asymmetry. As shown in Eq. (2.18), it is desirable to minimize the mechanical imperfections within the mechanical element of the gyroscope, so that their effect on precession pattern is minimal. However, it is also understood that for any real mechanical element, no matter how well it is built, there will be a certain degree of structural asymmetry and damping. This nonzero damping within the structure would result in the amplitude of oscillation or stored energy (E) of the gyroscope to decay over time and the unwanted quadrature error (Q) to grow due to aniso-elasticity. For this reason, if continuous RIG operation is desired, the residual structural asymmetry and damping need to be compensated through a closed loop control system. Such a closed-loop control system would need to fulfill the following roles:

1) Orientation of the precession pattern θ needs to be estimated, so that the magnitude of the rotation input can be measured, which is the gyroscope output.
2) Energy needs to be continuously added along the orientation of the precession pattern (θ), so that the amplitude of oscillation (E) remains constant.
3) Quadrature error (Q) needs to be nulled so that:
    - The vibration pattern never degenerates into the critical condition $a = q$,
    - The quadrature error does not interfere with estimation of the precession pattern.
4) Individual control should not interfere with the precession pattern while accomplishing their respective functions.

This functionality is typically achieved through the following control loops [22]:

1) A Phase Locked Loop (PLL), so that the gyroscope output can be demodulated and the pattern angle can be estimated along with other canonical variables E and Q,
2) An Amplitude Gain Control (AGC) loop to keep the amplitude of oscillation (E) stable,
3) A quadrature null loop to drive the quadrature error (Q) to zero.

In addition, if rate gyroscope operation is desired, an optional FTR loop can be utilized to control the orientation of the precession pattern to achieve the rate gyroscope mechanization.

3.3.1 Control System Overview

The control system is responsible for driving the gyroscope to a constant amplitude, suppressing the quadrature error, tracking/controlling the orientation of the precession pattern, without interfering with the measured precession pattern.

The key component of this approach is a PLL that tracks the gyro motion at any arbitrary pattern angle as opposed to locking to one of the drive axis (as in the open-loop rate gyroscope mechanization), Figure 3.4. PLL is used to demodulate in-phase and in-quadrature signals from each of the two gyro channels. From these

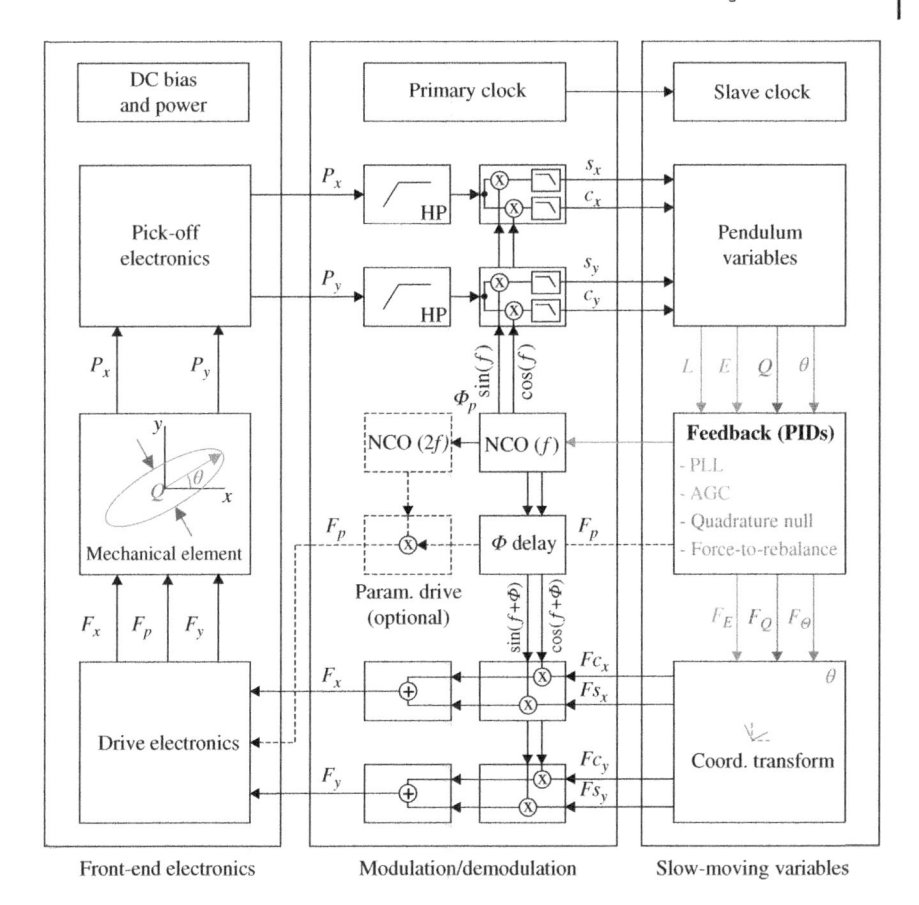

Figure 3.4 Whole-angle gyroscope control with (optional) parametric drive.

demodulated signals, the slow moving variables amplitude (E), quadrature error (Q), and pattern angle (θ) can be extracted using the equations described below [4]:

$$E = c_x^2 + s_x^2 + c_y^2 + s_y^2, \tag{3.1}$$

$$Q = 2(c_x s_y - c_y s_x), \tag{3.2}$$

$$R = c_x^2 + s_x^2 - c_y^2 - s_y^2, \tag{3.3}$$

$$S = 2(c_x c_y + s_x s_y), \tag{3.4}$$

$$L = c_x^2 - s_x^2 + c_y^2 - s_y^2 + 2i(c_x s_x + c_y s_y), \tag{3.5}$$

where E is a measure of energy within the system and is used for amplitude stabilization; Q is the measure of quadrature error and is independent of the drive orientation.

The imaginary component of L is a measure of the phase error and is used to establish a phase lock to the vibratory motion of the gyroscope. R and S are the projections of pattern angle on the x and y axis and can be used to find the orientation of the precession pattern angle using:

$$\theta = \frac{1}{2}\arctan\left(\frac{S}{R}\right). \tag{3.6}$$

A PID controller acts on each of these variables. These are the AGC acting on E, quadrature null acting on Q, and FTR that controls the pattern angle (θ). For the actual whole-angle operation, FTR loop is disabled so that the standing wave is free to precess. Once the correct command voltages F_E, F_Q, and F_θ are established, a coordinate transform around θ is performed to align these signals to the standing wave pattern:

$$F_{c_x} = F_E \cdot \cos(\theta) - F_\theta \cdot \sin(\theta), \tag{3.7}$$

$$F_{c_y} = F_E \cdot \sin(\theta) + F_\theta \cdot \cos(\theta), \tag{3.8}$$

$$F_{s_x} = -F_Q \cdot \sin(\theta), \tag{3.9}$$

$$F_{s_y} = F_Q \cdot \cos(\theta). \tag{3.10}$$

This is followed by modulation of the command voltages at the PLL frequency. A set amount of phase delay is also added during modulation, so that the total phase of the feed-back system is a multiple of 360°.

3.3.2 Amplitude Gain Control

Amplitude Gain Control (AGC) of whole-angle gyroscopes can be achieved via two different control architectures: (i) vector drive and (ii) parametric drive. Both of these architectures rely on a PID loop to control the amplitude of the parameter E. The main difference between the two architectures lies in the excitation scheme being used.

3.3.2.1 Vector Drive
In the vector drive, the amplitude control requires finding the orientation of the standing wave pattern and pumping energy along this direction using two drive channels, Eq. (3.10). Due to the dependence on the orientation of the standing wave pattern, this method is susceptible to drift due to gain unbalance in drive electronics, errors in calculating the angle of the standing wave, as well as the time delay between estimation of the angle of the standing wave pattern and the actual amplitude command.

3.3.2.2 Parametric Drive

In the parametric drive, a single drive channel is used for amplitude control of both modes (scalar drive). Even though a single drive channel is used, due to the parametric pumping effect, the energy added to each (x and y) mode is proportional to the existing amplitude of the respective mode. This permits the amplitude control of the standing wave at any arbitrary angle with minimal amount of perturbation. The scalar nature of the amplitude controller helps to bypass errors associated with finding the orientation of the standing wave, time delay in the calculation, and x–y drive gain drift.

To implement the parametric drive, a drive signal at twice of the PLL frequency is generated. This signal is applied to the so-called parametric drive electrode to parametrically pump energy in the x–y plane at twice the resonance frequency, Figure 3.4. This is made possible by leveraging the electrostatic spring softening effect associated with nonlinearity of the parallel plate transducers, which creates a relationship between the voltage difference across electrodes and the resonance frequency of the primary modes according to:

$$k_{\text{electrical}} = -2 \cdot \frac{C_0 \cdot \Delta V^2}{g^2}, \tag{3.11}$$

$$\omega = \sqrt{\frac{k_{\text{mechanical}} - k_{\text{electrical}}}{m}}, \tag{3.12}$$

where $k_{\text{mechanical}}$ and $k_{\text{electrical}}$ are the mechanical and electrical stiffnesses, respectively, m is the mass, ΔV is the voltage difference across the electrode, g is the electrode gap, and ω is the resonance frequency.

Due to the parametric pumping effect, the energy added to each (x and y) mode is proportional to the existing amplitude of the respective mode [22]:

$$\ddot{x} + \frac{\omega_x}{Q_x}\dot{x} + \left(\omega_x^2 + \frac{F_p}{m_{\text{eq}}}\sin(2\omega t + \phi_p)\right)x = \frac{F_x}{m_{\text{eq}}}\sin(\omega t + \phi_f) + 2\eta\dot{y}\Omega_z, \tag{3.13}$$

$$\ddot{y} + \frac{\omega_y}{Q_y}\dot{y} + \left(\omega_y^2 + \frac{F_p}{m_{\text{eq}}}\sin(2\omega t + \phi_p)\right)y = \frac{F_y}{m_{\text{eq}}}\sin(\omega t + \phi_f) - 2\eta\dot{x}\Omega_z. \tag{3.14}$$

As previously defined, ω_x, ω_y are the resonance frequencies, Q_x, Q_y are the Q-factors of the two degenerate modes, m_{eq} is the equivalent mass of the vibratory system, η is the angular gain factor, ω is the drive frequency, and ϕ_p, ϕ_f are the phases of the parametric and the vector drives, respectively.

This creates a preferential direction of pumping along the orientation of the standing wave without relying on angle of the precession pattern θ:

$$\theta = \arctan\left(\frac{y}{x}\right). \tag{3.15}$$

The open-loop parametric drive is typically unstable for nominal drive amplitudes [23, 24], which causes the gyro amplitude to increase exponentially for a fixed parametric drive signal. For this reason, an AGC loop is required to control the parametric drive voltage as to keep the gyro amplitude stable. This closed-loop operation permits the parametric drive of the gyro at a wide range of drive amplitudes, outside of the stability boundary of the open-loop parametric drive.

Due to the nonlinearity of the parametric drive, at an initial startup, the vector drive can be used to stabilized the drive oscillation. Once the PLL and AGC have stabilized, the vector drive can be disabled and the parametric drive AGC can be enabled.

An additional benefit of the parametric drive for MEMS gyroscopes is the minimization of the electrical feed-through between the drive and sense channels [23]. Since the drive and sense signals both contain information about the orientation pattern at the resonance frequency of the gyroscope, feed-through between the drive and sense electronics has a detrimental effect on the sense signal. The parametric drive mitigates this problem by separating the frequency of the drive and pick-off channels. Since the parametric drive frequency is a multiple of systems drive frequency, the electrostatic feed-through into the sense channel can be filtered out.

3.3.3 Quadrature Null Loop

Similar to the AGC loop, the quadrature null loop can be implemented using two different architectures: (i) AC quadrature null and (ii) DC quadrature null. Both architectures utilize a PID loop to zero the quadrature displacement Q, the difference lies in the excitation scheme used.

3.3.3.1 AC Quadrature Null
The AC quadrature null architecture utilizes a PID loop and two sets of forcers to cancel the quadrature displacement by applying forces orthogonal to the angle of the precession pattern at the drive frequency, Eq. (3.10). Since the quadrature signal is separated from the Coriolis signal via a 90° phase shift, any error in the quadrature phase can result in perturbation of the Coriolis signal that is being measured.

3.3.3.2 DC Quadrature Null
The DC quadrature null architecture utilizes the so-called DC quadrature null electrodes that can apply forces to both the x and y modes simultaneously. By controlling the DC bias voltage across the DC quadrature null electrodes, the quadrature displacement can be nulled. Despite the additional complexity due to the

addition of the DC quadrature null electrodes, this is the preferred method for quadrature cancellation due to the fact that the phase tuning step and associated errors related to the AC quadrature null scheme can be bypassed.

3.3.4 Force-to-rebalance and Virtual Carouseling

Implementation of the FTR loop for the whole-angle gyroscopes is very similar to the one described for FTR gyroscopes operating in Rate Mode (Section 3.2.2), with a few key differences: (i) FTR loop can act at any arbitrary pattern angle and as a result enables the control of the precession pattern independent of the rate input, (ii) FTR loop is not enabled during the actual whole-angle operation.

The main use cases for FTR loop in whole-angle gyroscopes is to initialize the orientation of the precession pattern to a known angle at startup and to control the pattern angle during operation. The ability to control the pattern angle during operation opens the path for parameter identification and subsequent self-calibration via "Virtual Carouseling." This is achieved by sweeping the pattern angle via FTR loop and keeping track of parameters such as the applied drive force, quadrature null signal, PLL frequency, and other parameters against the pattern angle.

3.4 Conclusions

Three different strategies for CVG mechanization were reviewed in this chapter: open-loop, FTR, and whole-angle mechanizations. These control strategies will be demonstrated in Chapter 5 using prototype MEMS gyroscopes for illustration.

Despite fundamental differences between the control strategies, a closer look makes certain commonalities apparent. For example, AGC and PLL subsystems are used in all three control strategies with only small differences in implementation. Similarly, Quadrature and Coriolis duality is common to all three control strategies; as a result, either an open- or closed-loop quadrature cancellation scheme is used in all three mechanizations. In addition, it is possible to use multiple control strategies on the same type of mechanical element. For example, degenerate mode gyroscopes can be instrumented for rate measurement via FTR mechanization or can be instrumented for angle measurement via whole-angle mechanization. Whereas, mode-mismatched gyroscopes with a small frequency separation in their fundamental modes can be instrumented with both open-loop and FTR mechanizations.

There is no CVG control strategy that is best for all applications. The decision on which control strategy to use depends on the requirements and constraints of the intended application. For example, due to the simplicity and ease of implementation, open-loop mechanization is typically employed on consumer and automotive gyroscopes, where certain trade-offs in performance may be acceptable. On the other hand, FTR mechanization is typically used in industrial applications to achieve increased sensor bandwidth, improved linearity, and a better scale factor stability on mode-match gyroscopes. Finally, whole-angle mechanization is implemented on degenerate mode gyroscopes with a high degree of symmetry and is aimed at most demanding industrial and aerospace applications, where increased cost and complexity can be justified.

Part II

2-D Micro-Machined Whole-Angle Gyroscope Architectures

4

Overview of 2-D Micro-Machined Whole-Angle Gyroscopes

State-of-the-art in development of the 2-D micro-machined whole-angle gyroscopes are reviewed in this section. Due to the degenerate mode operation, fundamentally, all devices presented in this chapter are capable of the whole-angle gyroscope operation.

4.1 2-D Micro-Machined Whole-Angle Gyroscope Architectures

4.1.1 Lumped Mass Systems

Lumped mass systems closely resemble micro-machined gyroscopes, they are composed of one or more proof masses, shuttle assemblies, and arrays of parallel plates or comb fingers for actuation and detection. In multi-mass systems, weak springs and lever mechanisms can be included in order to synchronize different proof masses. The designs are typically x–y symmetric, even though axisymmetric variants do exist.

- A lumped mass micro-machined angle-measuring gyroscope was proposed in [25]. The gyroscope consists of drive and sense electrodes and a proof mass coupled to an isotropic suspension system, such that the proof mass can move in any direction. A central anchor architecture as well as an outer anchor architecture with a distributed suspension system were discussed. The gyroscope was later fabricatated using surface-micro-machining and a method for suppressing structural errors was proposed in [16, 26]. Open-loop free angle precession of the mechanical element was also demonstrated for up to 1 s duration.
- An amplitude amplified dual mass gyroscope was demonstrated in [27]. This architecture utilizes a 4 degrees-of-freedom mechanical element that consists of an inner (drive) proof mass and an outer (sense) proof mass. The coupling between the drive mass and sense mass allows amplification of the sense mode

displacement, providing an improvement in signal-to-noise ratio. An ARW of $0.0096°/\sqrt{h}$ and in-run bias stability of $0.09°/h$ have been demonstrated using this architecture.
- A lumped mass gyroscope architecture with concentrated spring suspensions was reported in [28]. By concentrating a majority of the critical suspension elements in a small area, the effects of fabrication imperfections was mitigated. A sub-Hz frequency split (Δf) and a 50 mHz frequency stability over 130°C temperature range was demonstrated.
- A Quadruple Mass Gyroscope (QMG) was reported in [29]. The gyroscope consisted of four tines coupled to each other using lever mechanisms to achieve phase synchronization. Q-factors over two million were reported in [8]. Allan Variance analysis using temperature compensation showed the Angle Random Walk (ARW) of $0.07°/\sqrt{h}$ and an in-run bias stability of $0.22°/h$ [30]. Open-loop rate integrating operation (free precession of pattern angle) on QMG was initially demonstrated in [31], followed by the demonstration of a closed-loop rate integrating operation in [32–34].
- A large-displacement lumped mass gyroscope was reported in [35]. Due to the shaped-comb finger design displacements as high as 12.9μm was reported. Q-factors were on the order of 25k [35] for a 9.5 kHz gyroscope. Allan Variance analysis for rate gyroscope operation showed an ARW of $0.78°/\sqrt{h}$ and an in-run bias stability of $6.1°/h$. Stress effects on gyroscope performance were investigated in [36, 37].

4.1.2 Ring/Disk Systems

Axisymmetric silicon MEMS gyroscopes typically take the form of extruded 2-D geometries such as rings, concentric ring systems, and disks. The most common fabrication method is Deep Reactive Etching (DRIE) of crystalline silicon, although polysilicon and even fused silica variants do exist.

4.1.2.1 Ring Gyroscopes

One of the earliest examples of MEMS ring gyros was reported by British Aerospace [38]. The gyroscope consists of a single ring held in place by outer suspension elements, Figure 1.2. Like most other gyroscopes in this category, the $n = 2$ wineglass modes of the ring structure are used for the gyroscope operation. One of the main advantages of the architecture is that the electrodes are placed both inside and outside the ring structure, effectively doubling the total capacitance of the gyroscope. These electrodes can be used for forcing, pick-off, as well as DC frequency tuning [39]. Even though originally it was marketed as a rate gyroscope, the rate integrating operation has also been demonstrated on the same type of architecture [40, 41].

Another MEMS ring gyroscope was reported in [42–44]. In contrast to the outer suspension elements seen in [38], this design utilized an inner suspension system and a central anchor structure to hold the ring in place. The gyroscope was fabricated using a high aspect ratio combined polysilicon and single-crystal silicon process (HARPSS) [45, 46]. HARPSS process combines surface and bulk micro-machining techniques to create high aspect ratio vibratory structures with very small capacitive gaps (as low as 50 nm). The vibratory structure is created by depositing a sacrificial SiO_2 layer on the side walls of DRIE-etched trenches, filling them with LPCVD polysilicon and finally etching away the sacrificial SiO_2 layer. The Q-factor of 1200 and an as-fabricated frequency split (Δf) of 63 Hz at 28 kHz was reported on a ring structure with 1.1 mm diameter. Rate gyroscope operation was demonstrated with a minimum detectable signal of 0.04°/s at 10 Hz bandwidth, limited by the noise in the interface electronics. The device is fundamentally compatible with the whole-angle mechanization.

4.1.2.2 Concentric Ring Systems

A Disk Resonator Gyroscope (DRG) was reported in [47–49]. The gyroscope consists of multiple concentric ring structures, connected to each other using spokes and a central anchor system. Concentric rings increase the modal mass of the gyroscope, whereas the central anchor decouples the vibratory system from the anchor. A Q-factor of 80 000 was measured at 14 kHz center frequency for a 8 mm DRG. The Allan Variance analysis of the rate gyroscope operation showed an ARW of $0.0023°/\sqrt{h}$ and an uncompensated in-run bias stability of 0.025–0.012°/h was reported. With temperature compensation, the in-run bias stability was further reduced to below 0.01°/h. The gyroscope is also potentially capable of whole-angle mechanization. A noise analysis of DRG and the closed-loop vibratory rate gyroscopes was reported in [50, 51].

A model-based approach for permanent frequency tuning of Disk Resonator Gyro was reported in [52]. Gold balls were used to iteratively tune the frequency splits of $n = 2$ and $n = 3$ wineglass modes of the gyroscope. A DRG was successfully tuned, starting from a frequency split of 14.1 Hz on the $n = 2$ mode, down to 0.07 Hz. A similar mass perturbation technique was also used to decouple the resonator from linear acceleration [53]. More recently, a wafer level frequency tuning process based on ablation of a Parylene protective layer was reported [65]. These techniques are critical for achieving the required symmetry level, for subsequent mechanization of devices as whole-angle gyroscopes.

Significantly smaller polysilicon and single crystal silicon DRG prototypes at 600 µm were later fabricated in epitaxial polysilicon encapsulation (EpiSeal) process. The EpiSeal process utilizes an ultra-clean high temperature encapsulation step to seal the device under high vacuum [55]. Single crystal silicon variants were geometrically compensated to achieve lower as-fabricated frequency splits (Δf).

Polysilicon DRG showed Q-factor around 60 000 and an as-fabricated frequency split (Δf) of 135 Hz [56] at ~250 kHz center frequency, whereas single crystal silicon DRG showed Q-factor around 100 000 [57] and as-fabricated frequency split (Δf) of 96 Hz at a center frequency of ~250 kHz. Allan Variance analysis of the rate gyro operation showed ARW as low as $0.12°/\sqrt{h}$ and an in-run bias stability as low as 1.5°/h [58]. Rate Integrating Gyroscope (RIG) operation was also demonstrated in [59].

A single-crystal-silicon Cylindrical Rate Integrating Gyroscope (CING) was reported in [60, 61]. The gyroscope was fabricated on a glass substrate and consists of concentric silicon cylinders (rings) that are connected to each other through a silicon back-plate. Electrostatic characterization of a 2.5 mm radius CING showed Q-factors on the order of 20 000 and an as-fabricated frequency split (Δf) of ~ 10 Hz at 18 kHz center frequency. A CING with 12 mm diameter was later reported with Q-factors up to ~ 100 000. On this 12 mm prototype, an ARW of $0.09°/\sqrt{h}$ and an in-run bias stability of 129°/h was demonstrated. Rate integrating operation has also been demonstrated in [62–64]. The main advantage of this architecture is an order of magnitude increase in modal mass over a single ring structure; however, it was later found that the gyroscope has an angular gain of 0.011 due to the fact that majority of the kinetic energy is stored in the out-of-plane mode [65].

4.1.2.3 Disk Gyroscopes

Bulk Acoustic Wave (BAW) silicon disk gyroscopes were reported in [66–69]. The gyroscope consists of a silicon disk with release holes held in place by a central anchor. As opposed to most flexural type gyroscopes the BAW disk gyroscopes operate in the MHz range. Lower amplitude of motion due to the higher stiffness of the vibratory modes is compensated by the use of HARPSS process [46] to create extremely small capacitive gaps of 180 nm. Q-factors as high as 243 000 were reported on higher order wineglass modes of a 800 μm BAW disk gyroscope at a center frequency of 5.9 MHz. Rate gyro operation of a 1200 μm BAW disk gyroscope showed an ARW of $0.28°/\sqrt{h}$ and an in-run bias stability of 17°/h. Due to the axisymmetric resonator element, the whole-angle mechanization is also fundamentally possible on this gyroscope.

4.2 2-D Micro-Machining Processes

In this section, common 2-D micro-machining processes for fabrication of whole-angle gyroscopes are presented.

4.2.1 Traditional Silicon MEMS Process

The traditional silicon MEMS gyroscope fabrication process uses Deep Reactive Ion Etching (DRIE) to define the mechanical element of the gyroscope. The process starts by creation of an interposer wafer. This wafer typically contains multiple layers for electrical routing. This is followed by bonding of the device layer to the interposer wafer and DRIE to define the mechanical element. Finally, glass frit bonding is used to create a hermetic seal between the cap layer and the device layer. Typically, a getter layer is also deposited into the cap layer to achieve higher levels of vacuum, Figure 4.1.

Once the fabrication is complete, the individual MEMS dies are singulated and integrated with electronics by wire-bonding to the CMOS dies.

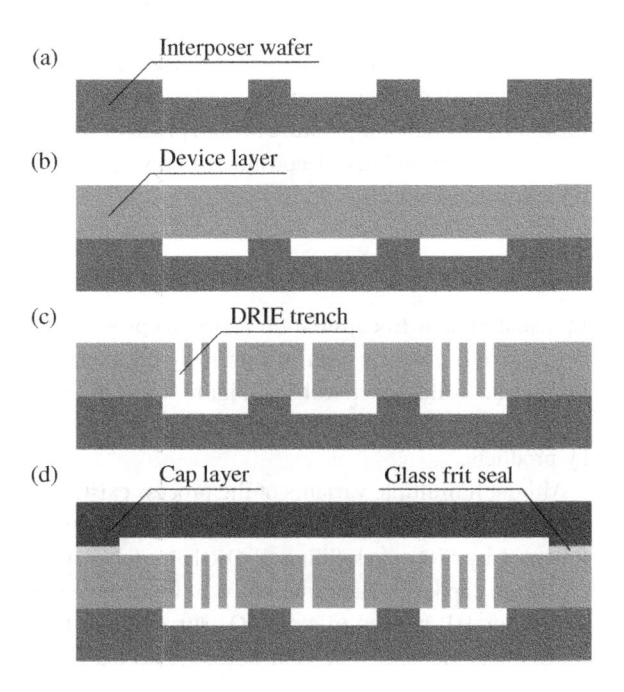

4.2.2 Integrated MEMS/CMOS Fabrication Process

By bonding the cap layer at the beginning of the process and using a low temperature eutectic bonding process at the end, it is possible to integrate CMOS and MEMS wafers in the same process [70]. Advantage of such a process over traditional wafer-level packaging process is the close integration of CMOS and MEMS components, eliminating the need for subsequent wire-bonding step and potentially reducing parasitic elements in front-end electronics. The process starts by

Figure 4.1 Traditional silicon MEMS encapsulation process consists of: (a) fabrication of interposer wafer, (b) bonding of device layer, (c) Deep Reactive Ion Etching (DRIE) of the device layer, and (d) glass frit bonding of the cap layer to create a hermetic seal.

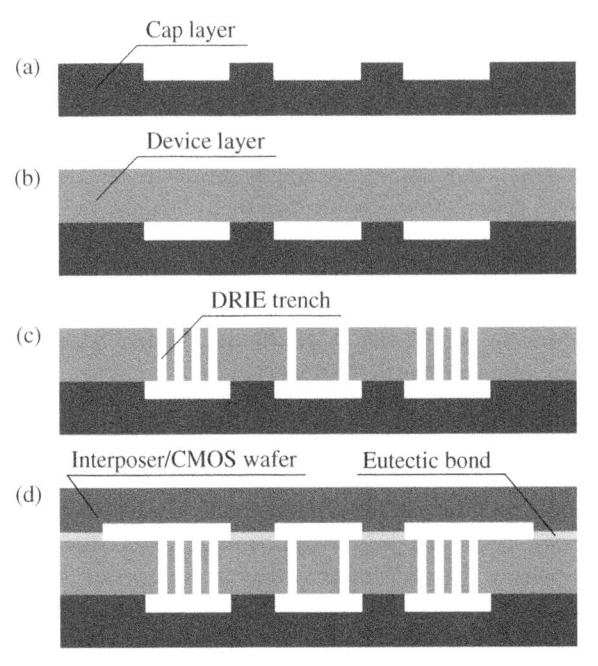

Figure 4.2 Integrated MEMS/CMOS fabrication process consists of: (a) pre-etching of the cap layer, (b) bonding of device layer, (c) Deep Reactive Ion Etching (DRIE) of the device layer, and (d) low temperature eutectic bonding of the CMOS wafer.

bonding the device layer to a pre-etched cap layer, followed by DRIE of the device layer. The final step of the process is bonding of the CMOS with the MEMS wafer, creating a hermetic seal and providing electrical connectivity in one step. By fabricating the CMOS wafer separately and using low temperature eutectic bonding as the final step, high temperature MEMS processes, such as fusion bonding and Low Pressure Chemical Vapor Deposition (LPCVD), can be used without damaging the CMOS wafer, Figure 4.2.

4.2.3 Epitaxial Silicon Encapsulation Process

Epitaxial Silicon Encapsulation (EpiSeal) process utilizes epitaxially grown silicon to seal the device layer at extremely high temperatures, which results in an ultra-clean wafer-level seal [7], Figure 4.3. This results in high vacuum levels, as low as 1 Pa, without the need for getter materials for absorption of sealing by-products.

Although multiple variants of the process exist, the fundamental process starts with DRIE of a conventional Silicon-on-Insulator (SOI) wafer stack. This is followed by SiO_2 back-filling of DRIE trenches and epitaxial silicon growth to create the cap layer. Subsequently, vent holes are etched in the cap layer via DRIE and vapor HF is used to etch SiO_2 surrounding the device layer to release the mechanical element. The vent holes are sealed using a second epitaxial growth

Figure 4.3 Epitaxial Silicon Encapsulation (EpiSeal) process consists of: (a) preparation of a Silicon-on-Insulator (SOI) wafer stack, (b) Deep Reactive Ion Etching (DRIE) of the device layer, (c) SiO_2 back-filling of the trenches, (d) epitaxial silicon growth, (e) DRIE of vent holes in cap layer, (f) vapor HF release of SiO_2, (g) sealing of the vent holes via further epitaxial silicon growth, and (h) creation of through silicon vias for electrical connectivity.

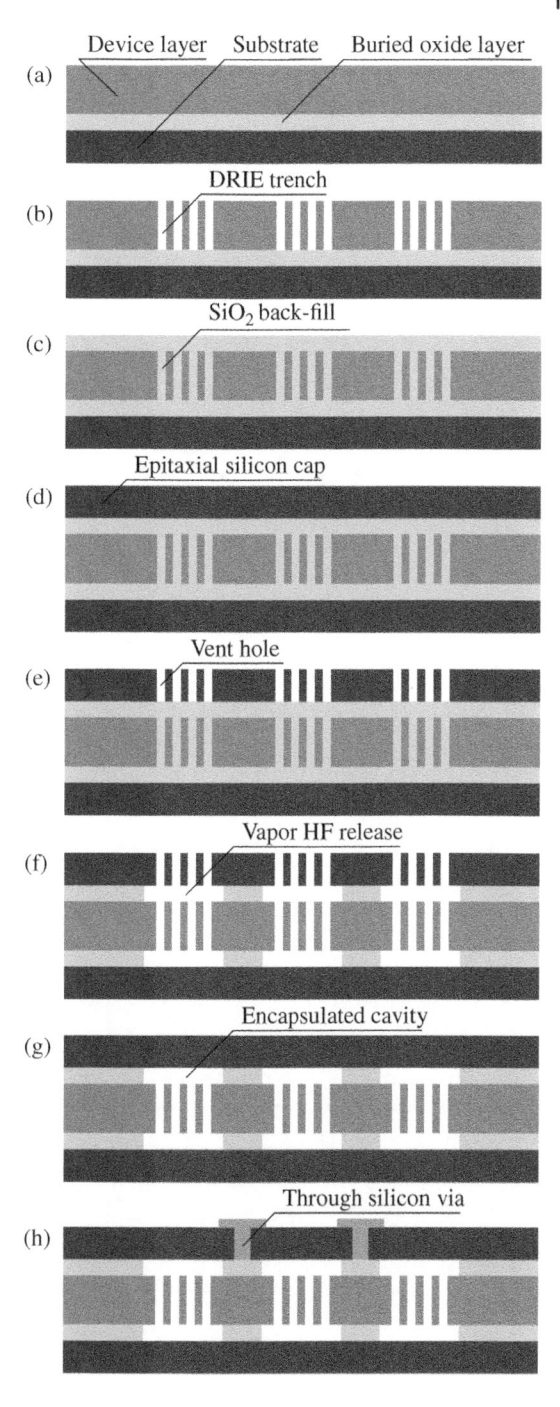

process. Finally, through silicon vias are created in the cap layer to provide electrical connectivity to the device layer.

The EpiSeal encapsulation process was proposed by researchers at the Robert Bosch Research and Technology Center in Palo Alto and then demonstrated in a close collaboration with Stanford University. This collaboration is continuing to develop improvements and extensions to this process for many applications, while the baseline process has been brought into commercial production by SiTime Inc., which was later acquired by MegaChips Corporation in 2014.

5

Example 2-D Micro-Machined Whole-Angle Gyroscopes

In this chapter, we review two examples of 2-D micro-machined whole-angle gyroscope architectures in order to illustrate factors that influence design decisions, fabrication considerations, and characterization methodology. First a silicon ring gyroscope, Toroidal Ring Gyroscope (TRG), will be reviewed. This will be followed by an overview of a silicon lumped mass gyroscope, Dual Foucault Pendulum (DFP), Figure 1.2.

5.1 A Distributed Mass MEMS Gyroscope – Toroidal Ring Gyroscope

In this section, TRG is reviewed to provide an example for conventionally micro-machined, 2-D silicon ring whole-angle gyroscope architectures, Figure 1.2. Other examples of ring/disk gyroscopes are the Disk Resonator Gyroscope (DRG) [47, 48], ring gyroscopes [42], and Bulk Acoustic Wave (BAW) disk gyroscopes [66, 71]. TRG belongs to the family of concentric ring gyroscopes. The gyroscopes that belong to this family utilize multiple concentric rings that are connected by spokes for their vibratory element.

One of the earliest examples of concentric ring gyroscopes is DRG, which utilizes a central anchor structure and multiple concentric rings around the central anchor structure for its vibratory element. In TRG, the device is anchored at the outer-most perimeter using a distributed support structure for anchor loss minimization, and as a result the vibrational energy in the introduced design is concentrated toward the innermost ring. The distributed support structure prevents vibrational motion propagating to the outer anchor, which helps trap the vibrational energy within the gyroscope [72], Figure 5.1.

Whole-Angle MEMS Gyroscopes: Challenges and Opportunities,
First Edition. Doruk Senkal and Andrei M. Shkel.
© 2020 The Institute of Electrical and Electronics Engineers, Inc. Published 2020 by John Wiley & Sons, Inc.

48 | *5 Example 2-D Micro-Machined Whole-Angle Gyroscopes*

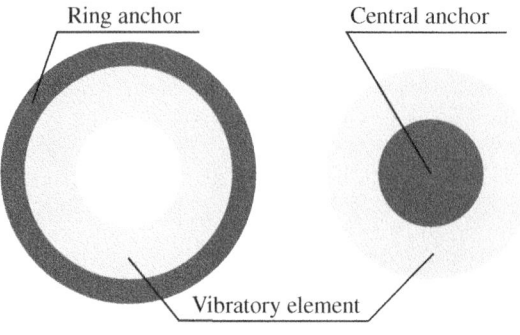

Figure 5.1 Ring/disk gyroscopes can be anchored (a) at the outer perimeter of the vibratory element, or (b) at the center-point of the vibratory element.

(a) Toroidal Ring Gyro (TRG)   (b) Disk/ring gyros

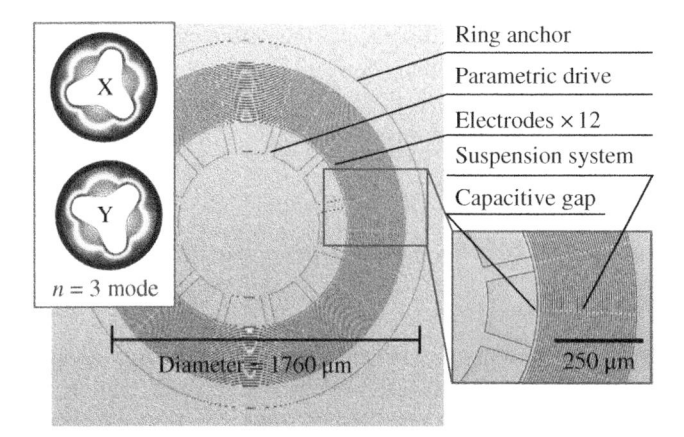

Figure 5.2 A 100k Q-factor, epitaxial silicon encapsulated Toroidal Ring Gyroscope was used in the experiments. Device consists of an outer ring anchor, distributed suspension system, and an inner electrode assembly. *Source:* SEM image from [74]. © IEEE.

5.1.1 Architecture

TRG consists of an outer anchor that encircles the device, a distributed ring support structure, and an inner electrode assembly, Figure 5.2. During oscillation, the vibration energy is concentrated at the innermost ring, Figure 5.4. The distributed support structure decouples the vibrational motion from the substrate, Figure 5.4. This decoupling mitigates anchor losses into the substrate and prevents die/package stresses from propagating into the vibratory structure.

In order to illustrate design considerations, fabrication challenges, and characterization methodology of 2-D silicon ring whole-angle gyroscope architectures, a prototype TRG was built. The device was fabricated on a 2 mm × 2 mm die, the mechanical element had an outer diameter of 1760 μm, and was fabricated on a

Table 5.1 Summary of design parameters.

Device diameter	(μm)	1760
Device layer thickness	(μm)	40
Capacitive gaps	(μm)	1.5
Ring thickness	(μm)	5
Innermost ring thickness	(μm)	8.5
Number of rings		44

device layer thickness of 40 μm, Table 5.1. In the illustrated design, the suspension system consists of 44 concentric rings. The rings are connected to each other using 12 spokes between the rings, the spokes are interleaved with an offset of 15° between two consecutive rings. The suspension rings have a thickness of 5 μm. The innermost ring is designed to have a slightly higher ring thickness of 8.5 μm; this mitigates the effect of spokes on the overall mode shape and helps retain a truer wineglass shape at the electrode interface.

In a later study on TRGs, it was shown that it is possible to re-order the wineglass modes by changing the number of rings and the angle between each spoke [73]. Mode ordering can be used to mitigate the unwanted effects of parasitic modes by making them stiffer and as a result improve vibration immunity of the sensor.

5.1.1.1 Electrode Architecture

In this example implementation, the electrode assembly is located at the center of the gyroscope and consists of 12 discrete electrodes and one central star electrode, Figure 5.3. Discrete electrodes are distributed in groups of six for the two degenerate wineglass modes. Out of six electrodes, two electrodes were used as a forcer and four as a pick-off for each mode, giving a total of four forcer and eight pick-off electrodes across the gyro.

In this illustration, the central star-shaped electrode is used for parametric pumping. Due to the dodecagram (12-pointed) star shape of this electrode, parametric pumping has equal contribution to the both degenerate $n = 3$ wineglass modes.

5.1.2 Experimental Demonstration of the Concept

5.1.2.1 Fabrication

The TRG was fabricated using a wafer-level Epitaxial Silicon Encapsulation (EpiSeal) process [7]. EpiSeal process utilizes epitaxially grown silicon to seal the device layer at extremely high temperatures, which results in an ultra-clean wafer-level seal. This results in high vacuum levels (as low as 1 Pa) without the need for getter materials for absorption of sealing by-products.

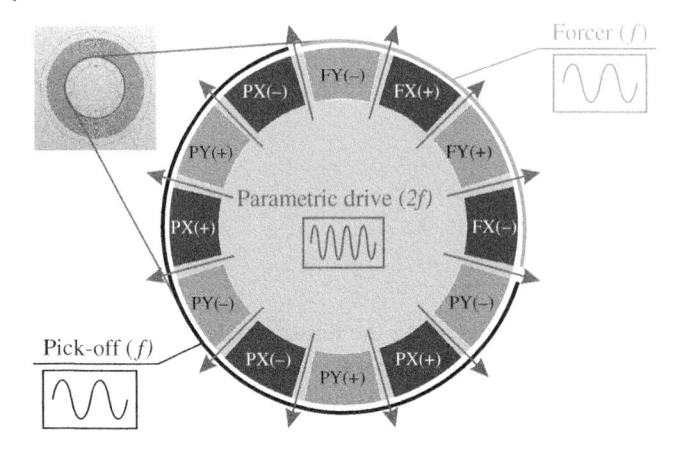

Figure 5.3 In this implementation, the central electrode assembly consists of 12 discrete electrodes, divided into 4 drive and 8 pick off electrodes, and 1 star-shaped parametric electrode.

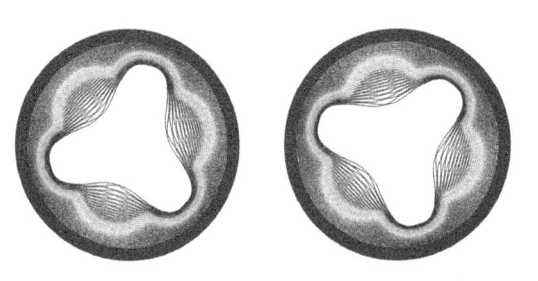

Figure 5.4 Due to the distributed suspension system vibrational energy is trapped within the structure.

Vibration amplitude (nm) for 200 nm peak displacement

| 0 | 10 | 20 | 30 | 40 | > 50 |

The device was fabricated on Epitaxial Silicon Encapsulation (EpiSeal) process [7] using single-crystal ⟨100⟩ device layer. For this reason, the device was designed to operate in $n = 3$ wineglass modes instead of the more commonly used lower order $n = 2$ wineglass modes, Figure 5.4. This approach eliminates frequency split induced on $n = 2$ wineglass modes by anisotropic modulus of elasticity of crystalline silicon and makes the frequency splits (Δf) insensitive to misalignment errors in crystalline orientation of the silicon wafer. The drawbacks of operation in $n = 3$ modes are slightly lower angular gain factor, higher resonance frequency, and smaller amplitude of motion.

5.1.2.2 Experimental Setup

Devices were wire-bonded to ceramic Leadless Chip Carrier (LCC) packages and instrumented with discrete electronics. Electromechanical Amplitude Modulation (EAM) using a carrier at 1 MHz was used to mitigate the effects

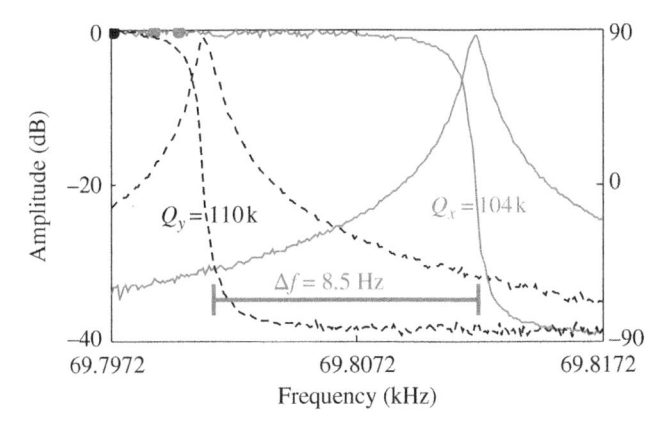

Figure 5.5 Frequency sweep showing the $n = 3$ wineglass modes with Q-factor above 100k at central frequency of 69.8 kHz.

of parasitic feed through on the pick-off electronics. DC bias voltage of 1 V on both modes and AC voltage of 20 mV was used for initial characterization. Frequency response characterization of the fabricated gyroscopes revealed a Q-factor of >100 000 on both $n = 3$ modes at ~70 kHz center frequency, Figure 5.5.

5.1.2.3 Mechanical Characterization

As-fabricated frequency split (Δf) of four devices were characterized. Typical for MEMS fabrication technology, the lowest frequency split observed was 8.5 Hz, Figure 5.5, ($\Delta f/f = 122$ ppm) with a mean frequency split of 21 Hz ($\Delta f/f = 300$ ppm) across four devices, Table 5.2. Low frequency split is attributed to robustness of the high order ($n = 3$) wineglass mode to fabrication imperfections and the ultra-clean EpiSeal process.

After initial characterization, the frequency split was further reduced using electrostatic tuning of DC bias on forcer and quadrature null electrodes. DC bias voltages of 3.26 and 0.5 V was sufficient to reduce the frequency split to <100 mHz ($\Delta f/f < 2$ ppm) on device 1, Figure 5.6.

Table 5.2 As-fabricated frequency symmetry of four devices.

Device no.	Δf(Hz)	f(kHz)	$\Delta f/f$(ppm)
1	8.5	69.75	122
2	11	69.69	158
3	25	71.29	350
4	40	69.4	576

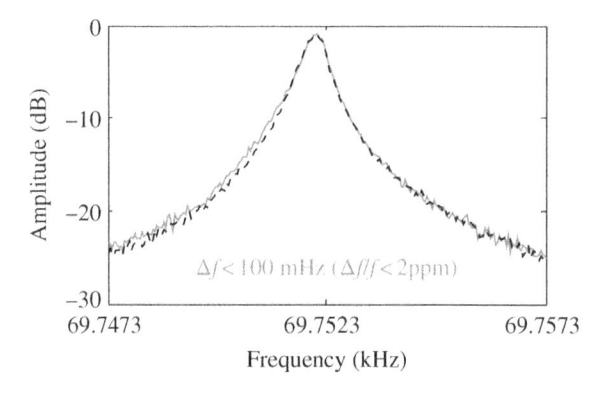

Figure 5.6 Electrostatic tuning with 3.26 and 0.5 V resulted in $\Delta f < 100$ mHz ($\Delta f < 2$ ppm at 69.75 kHz).

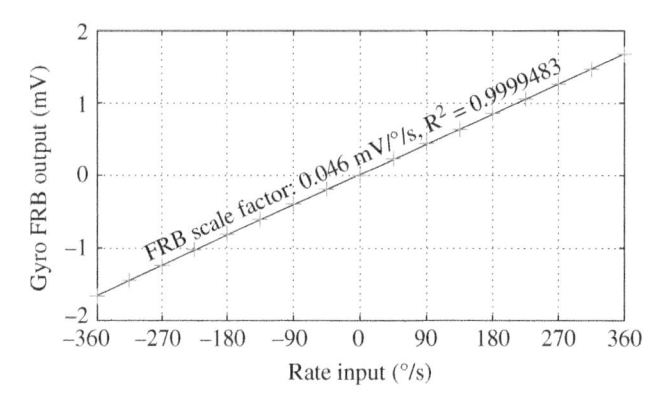

Figure 5.7 Scale factor of Toroidal Ring Gyroscope in force-to-rebalance mode.

For force-to-rebalance (FTR) rate gyroscope operation, an ADAU1442 DSP board from Analog Devices was used to implement the control algorithms. A 24 bit audio codec (AD1938) operating at 192 kHz sampling rate was used for forcer and pick-off signals. An Arduino Due micro-controller board was interfaced with the DSP board over I2S protocol, which was used to down-sample the gyro output and transmit over RS232 protocol for data acquisition on a PC.

5.1.2.4 Rate Gyroscope Operation

To evaluate the rate gyro performance, the precession pattern was locked to x-axis by enabling the FTR loop. A FTR scale factor of 0.046 m V($°/s$) was observed, Figure 5.7. Allan Variance analysis in force-rebalance operation without any

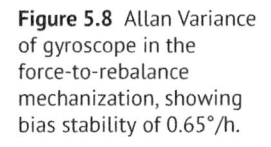

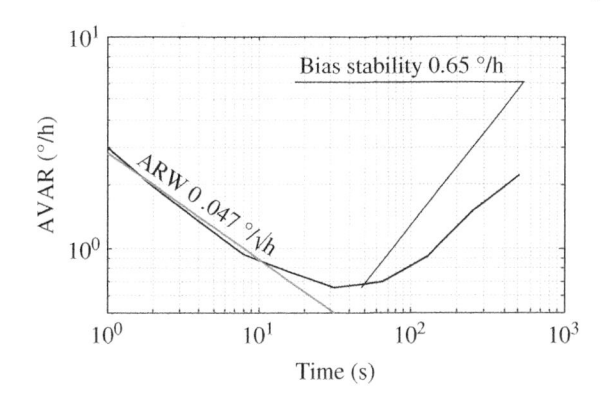

Figure 5.8 Allan Variance of gyroscope in the force-to-rebalance mechanization, showing bias stability of 0.65°/h.

temperature compensation or stabilization revealed an ARW of $0.047°/\sqrt{h}$ and an in-run bias stability of 0.65°/h at 32 s integration time, Figure 5.8.

5.1.2.5 Comparison of Vector Drive and Parametric Drive

In order to observe the impact of the parametric drive on whole-angle gyroscope performance, a TRG, was instrumented with the parametric drive. In this demonstration, a central star electrode was used for parametric pumping, which has equal contribution to the both the degenerate $n = 3$ wineglass modes, Figure 5.3. To test the parametric drive performance, a constant rotation rate was applied for 1 h, switching the direction at 30 min mark. Figure 5.9 shows the unwrapped gyro response for four different speeds. This experiment was later repeated using the conventional (vector) drive. A linear fit to the data revealed a combined electrical/mechanical angular gain factor of ~0.6. Comparison of residuals from both experiments is shown in Figure 5.10. For all rate inputs, the parametric drive resulted in smaller residual errors compared to conventional drive architecture. As expected, the highest difference between the conventional drive and the parametric drive occurred at higher rotation rates: time delay in the calculation of the pattern angle becomes more important at higher rotation rates as this delay can result in a coupling between the drive force and the FTR force. Any change in the drive force amplitude either due to change in the drive electronics gain or a Q-factor change in the resonator element can affect the scale factor stability.

For a 360 °/s rate input over a 30 min period, the standard deviation of accumulated error for the vector drive was 176° versus 13° for parametric drive, which corresponds to 14× improvement for parametric drive and a scale factor variation of <20 ppm STD.

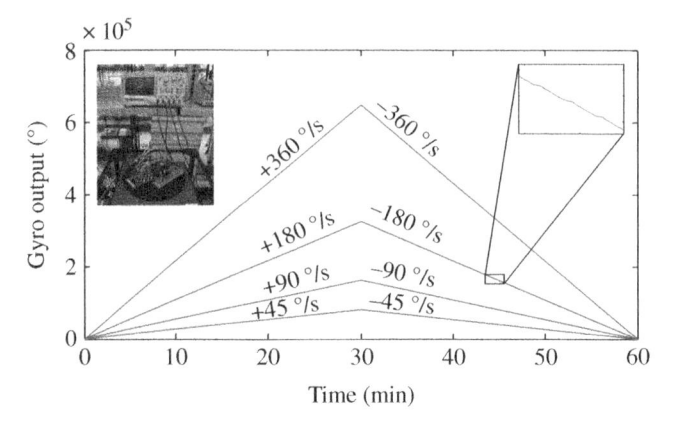

Figure 5.9 Experimental demonstration of rate integrating operation under parametric drive.

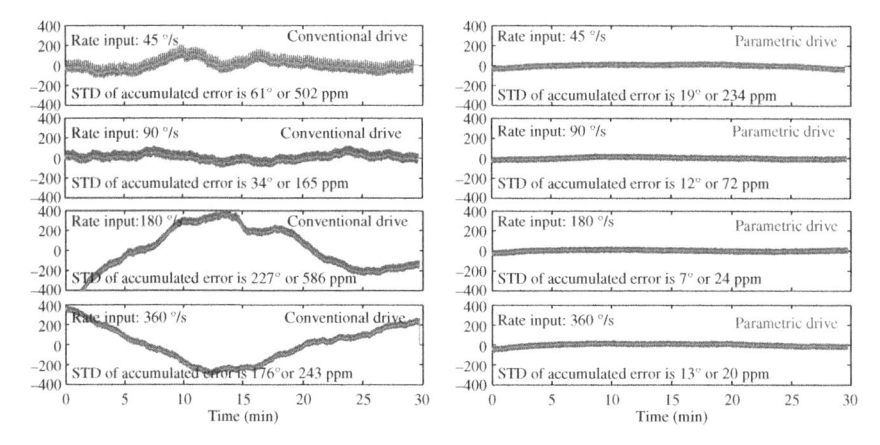

Figure 5.10 Comparison of residual errors of conventional drive and parametric drive for different rate inputs.

5.2 A Lumped Mass MEMS Gyroscope – Dual Foucault Pendulum Gyroscope

In this section, we review a conventionally micro-machined, lumped mass whole-angle gyroscope architecture: DFP gyroscope. DFP belongs to the family of lumped mass gyroscope architectures, Figure 1.2, and consists of two dynamically equivalent, mechanically coupled proof masses, oscillating in the anti-phase motion. This dual axis tuning fork behavior creates a dynamically

Figure 5.11 Wineglass modes of axisymmetric architectures, such as ring/disk systems and wineglasses gyroscopes, are inherently dynamically balanced, resulting in zero net reaction force and torque during oscillation.

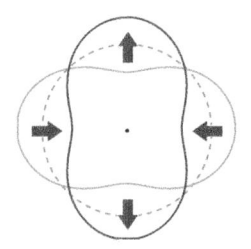

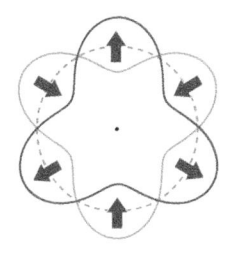

(a) $n = 2$ wineglass mode

(b) $n = 3$ wineglass mode

balanced resonator with x–y symmetry in frequency and damping, suitable for the whole-angle mechanization.

Axisymmetric gyroscope architectures that utilize the so-called "wineglass" modes, such as ring/disk systems and wineglass gyroscopes, are inherently force and torque balanced. The wineglass modes exhibit a behavior akin to a tuning fork, where the motion at one part of the gyroscope is balanced by an equal and opposite motion, Figure 5.11 [75, 76]. This balanced motion provides anchor loss mitigation and a certain degree of vibration isolation.

Conventional micro-machined gyroscopes, on the other hand, typically consist of proof masses, folded beam suspensions, and comb fingers, that are not inherently force and torque balanced. To achieve force and torque balance, tuning fork architectures such as that in [77] can be utilized. However, conventional tuning fork architectures are not x–y symmetric and the force and torque balance occurs along one of the axis. On the other hand, x–y symmetric single-mass systems, such as [35], or multi-mass systems such as the Quadruple Mass Gyroscope (QMG) [78], do exist. Single-mass systems are not inherently balanced, whereas multi-mass systems require complicated mechanical systems to achieve force/torque balance and phase synchronization.

DFP combines simplicity and dynamic balance of tuning fork gyros [77] (two proof masses in anti-phase motion) with high rate sensitivity of degenerate mode gyroscopes (x–y symmetry). The core of the gyroscope architecture is two mechanically coupled and dynamically equivalent proof masses, oscillating in anti-phase motion, Figure 5.12b. Each proof mass is free to swing in any direction on the x–y plane, analogous to a Foucault Pendulum, Figure 5.12a. However, unlike a conventional tuning fork gyroscope, the center of masses of the two proof masses are aligned. This creates force and moment balance for both x and y modes, providing immunity to vibration and shock as well as anchor loss mitigation. It is believed that this two-mass architecture is the minimum lumped mass gyroscope configuration that can provide a dynamically balanced system in both x and y directions.

A prototype DFP gyroscope was built in order to illustrate the design considerations, fabrication challenges, and characterization of 2-D silicon lumped mass whole-angle gyroscopes.

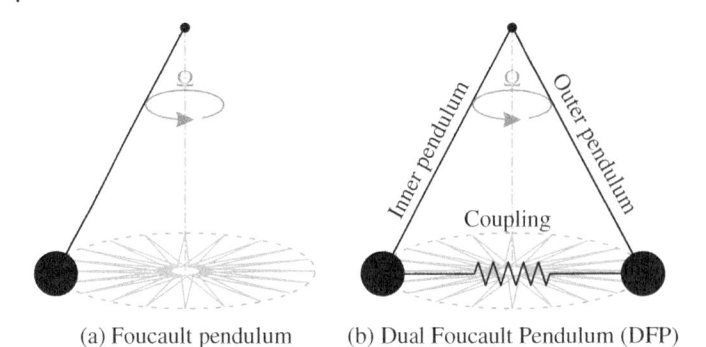

(a) Foucault pendulum   (b) Dual Foucault Pendulum (DFP)

Figure 5.12 Dual Foucault Pendulum (DFP) gyroscope consists of two mechanically coupled Foucault Pendulums.

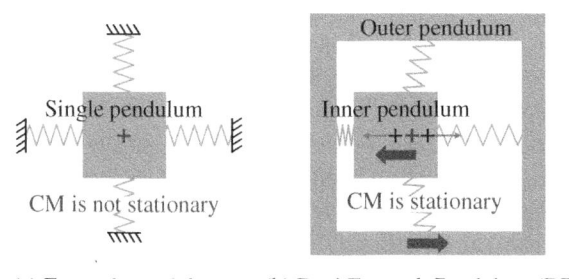

(a) Foucault pendulum   (b) Dual Foucault Pendulum (DFP)

Figure 5.13 Vibration immunity and anchor loss mitigation are provided by anti-phase operation of two dynamically equivalent Foucault pendulums.

5.2.1 Architecture

The core of the DFP gyroscope is two mechanically coupled and dynamically equivalent proof masses, oscillating in anti-phase motion. The dynamic balance is obtained by aligning the center of masses of each proof mass. This allows the center of mass of the system to remain stationary during oscillation, causing the net forces and torques generated by the combined system to be zero at all times, Figure 5.13. Unlike a conventional tuning fork gyroscope, the force and torque balance is obtained on both x and y modes of the gyroscope.

Dynamic equivalence of the two proof masses is achieved by using identical (mirrored) suspension elements and shuttle assemblies, while designing the two proof masses to have equal masses. This results in same resonance frequencies for individual proof masses, which is further reinforced by mechanical coupling of the two proof masses. This mechanical coupling is achieved via "weak springs" between shuttle assemblies of inner and outer proof masses, which synchronizes the phases of the proof masses, Figure 5.14.

(a) X-mode   (b) Y-mode

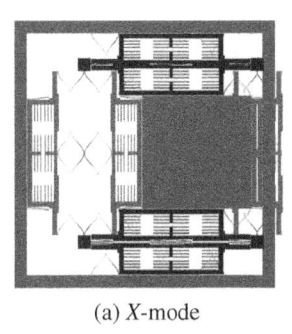

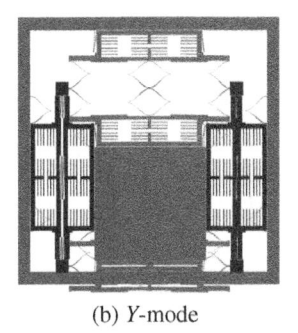

Figure 5.14 FEA showing x–y symmetric anti-phase operation. Device is anchored at four points in between the proof masses. Colors correspond to total displacement.

The device is suspended from four anchors placed in between the two proof masses. Each anchor is shared by one x and one y shuttle pair. To help protect the mechanical element from unwanted packaging stresses, attachment of the gyroscope die to the package is done in between the four anchors, via a central attachment point.

5.2.1.1 Electrode Architecture

There are four shuttle pairs within the gyroscope. Each shuttle pair is connected to both inner and outer proof masses and can only move in one direction. This helps to mitigate cross-axis coupling between the x and y modes by restricting electrode movement in one direction. During gyroscope operation, for each x and y mode, two shuttle pairs remain parked, whereas the other two shuttles oscillate in anti-phase motion together with their respective proof masses.

Electrostatic transduction is provided by arrays of parallel plates located on the shuttle assemblies. In order to achieve large displacements necessary for low noise operation, 8 µm capacitive gaps are used on the parallel plates. A large free space between the two proof masses allows placement of 12 layers of parallel plate electrodes per shuttle pair, resulting in over 12.5 pF total capacitance ($dC/dx = 1.5$ µF/m).

5.2.2 Experimental Demonstration of the Concept

5.2.2.1 Fabrication

A prototype of the device was fabricated in a standard SOI process, with a footprint of 6700 µm × 6700 µm, Figure 5.15. A device layer of 100 µm and a buried oxide layer of 5 µm were used. AZ 4620 photoresist and conventional contact lithography were used to define the sensor features. DRIE etching of the device layer was done in an STS DRIE system, which was followed by an HF release step using an Idonus

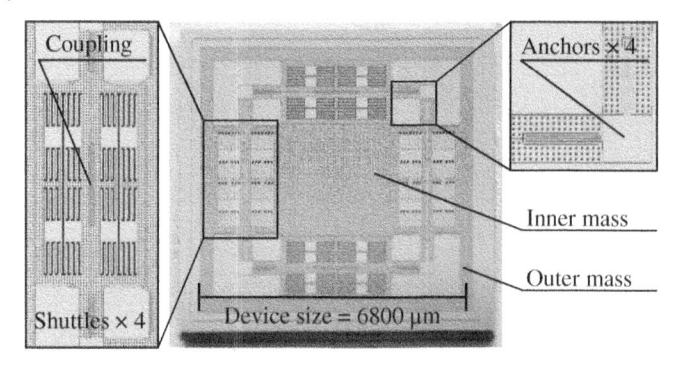

Figure 5.15 Image of fabricated gyroscope with closeups of the shuttle assemblies and the anchors.

Table 5.3 Summary of design parameters.

Device width and height	(µm)	6800
Device layer thickness	(µm)	100
Buried oxide (BOX) thickness	(µm)	5
Buried oxide (BOX) thickness	(µm)	5
Release hole width and height	(µm)	20
Inner tine width and height	(µm)	2880
Outer tine frame width and height	(µm)	320
Folded beam width	(µm)	10
Folded beam length	(µm)	550
Number of folded beams	#	48
Nominal center frequency	(kHz)	3
Parallel plate overlap length	#	1600
Number of parallel plates per axis	#	24
Capacitive gaps	(µm)	8
Stop gaps	(µm)	5
Total nominal capacitance	(pF)	12.5

Vapor Phase Etcher. After dicing, individual dies were attached to 44 pin ceramic LCC packages and wirebonded for characterization, Table 5.3.

5.2.2.2 Experimental Setup

A low-outgassing ceramic PCB was used for front-end electronics. The first stage amplification of the gyroscope output was done using dual trans-impedance amplifiers (Analog Devices AD8066) with 1 MΩ gain resistors and 2.2 pF

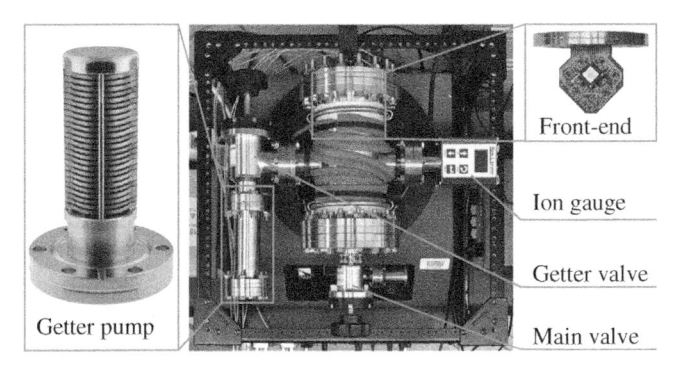

Figure 5.16 High-vacuum test-bed with nonevaporable getter pump provided μTorr level vacuum for rate characterization.

capacitors. Output of the transimpedance amplifiers were cascaded into an instrumentation amplifier (Analog Devices AD8429).

The same instrumentation amplifiers (Analog Devices AD8429) were also used for forcer electronics. Unity gain was used on the forcer electronics due to extremely low voltage levels required to drive the gyroscope (less than 1 mV). DC biasing was done only on the forcer electrodes and the resonator.

Finally, low dropout voltage regulators (Texas Instruments TPS7A3001 and TPS7A4901) were used for supplying power to the active components on the PCB. These helped to reduce system noise by rejecting a large portion of the power supply noise.

A high-vacuum test-bed was used for gyro characterization, Figure 5.16. The test-bed consists of four main components:

- Low-outgassing ceramic PCB front-end electronics,
- Macroscale nonevaporable getter pump,
- Stainless steel vacuum chamber assembly,
- Rate table with slip rings.

For the rate table characterization, the device was mounted onto the front-end PCB and inserted into the vacuum chamber assembly. Electrical feed-through from the vacuum chamber was supplied by a 37 pin D-SUB connector, which was then routed through the slip rings. An angle valve was used to seal the getter pump, during insertion of the device into the vacuum chamber. Another angle valve was used to seal the entire vacuum chamber, so that the external turbo pump can be disconnected for continuous 360° rotation of the rate table.

After the system was pumped down using an external turbo pump, the nonevaporable getter pump was activated using the internal resistive heater and the chamber was sealed off. Due to the large absorption capacity of the getter pump and the low-outgassing ceramic front-end electronics, the system could sustain high vacuum without the need for active pumping. This eliminates unwanted vibrations caused by rotary pump systems and permits continuous 360° rotation of the rate table at sustained vacuum levels of <10 μTorr.

5.2.2.3 Mechanical Characterization

Ring-down characterization was used to measure the Q-factor of the mechanical element. An exponential curve fit to the ring-down data showed an energy decay time constant (τ) of 30 s at 3.2 kHz, which corresponds to Q-factor over 300 000, Figure 5.17. A later study performed in [79] using getter sealed DFP gyroscopes, confirmed the viscous damping to be the dominant loss mechanism by demonstrating a Q-factor over 700 000.

An as-fabricated frequency split (Δf) of 18 Hz was observed, which was later electrostatically tuned to <100 mHz by biasing the forcer electrodes. This was achieved by applying a DC bias of 10 v DC at the resonator, while applying 9 V DC to the x forcer electrodes and −6.75 V DC to the y forcer electrodes. Lower DC bias voltages can be used if pick-off electrodes in addition to the forcer electrodes are used for electrostatic tuning.

5.2.2.4 Rate Gyroscope Operation

After electrostatic tuning, Phase Locked Loop (PLL), Amplitude Gain Control (AGC), and quadrature null loops were implemented on a Zurich HF2LI lock-in amplifier [22]. PLL was locked to the drive mode and AGC was used to stabilize the drive amplitude. In the experiments, an AC quadrature null loop was utilized, where the sense mode forcer electrodes were used to drive the quadrature output of the sense mode to zero. The device was tested with both whole-angle and FTR configurations.

Scale factor characterization was done using continuous rotation of the rate table and incrementally changing the angular velocity. A linear fit to the gyro output was used to extract the scale factor. An open-loop scale factor of 26.4 mV/(°/s) and FTR scale factor of 4.66 mV/(°/s) were measured, Figure 5.18. After the scale factor was obtained, Allan Variance analysis of the gyroscope zero rate output was performed for both whole-angle and FTR operation, Figure 5.19. No temperature stabilization or compensation was used during the experiment. For whole-angle operation, an Angle Random Walk (ARW) of $0.003°/\sqrt{h}$ and an in-run bias stability of $0.27°/h$ were measured. Whereas for FRB configuration an ARW of $0.06°/\sqrt{h}$ and a bias stability of $1.5°/h$ were measured. Higher ARW in FRB operation was associated with feedback noise from the digital to analog converters (DACs).

5.2.2.5 Parameter Identification

In order to demonstrate the Parameter Identification via "Virtual Carouseling," a DFP gyroscope was instrumented with the whole-angle mechanization, Figure 3.4. The pattern angle dependence of PLL, AGC, AC quadrature null,

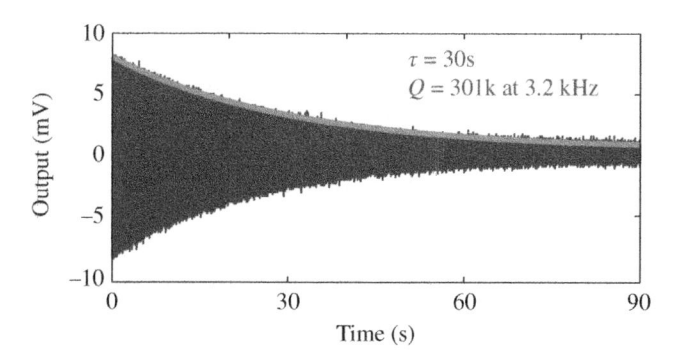

Figure 5.17 Ring-down experiment showing energy decay time constant (τ) of 30 s or Q of 301k at 3.2 kHz.

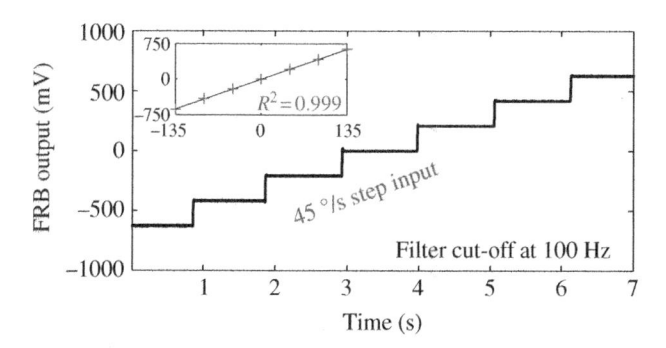

Figure 5.18 Rate characterization with 45°/s step input showed a FRB scale factor of 4.66 mV/(°/s), goodness of fit: $R^2 = 0.999$.

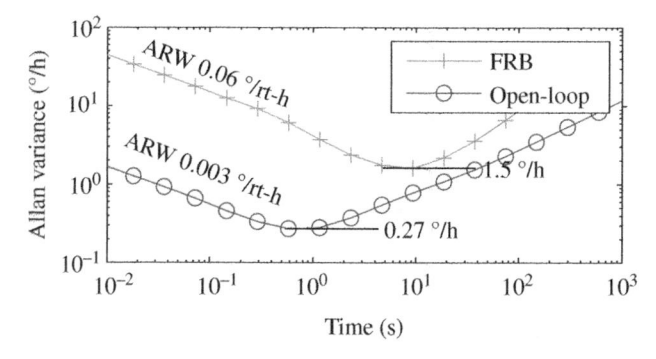

Figure 5.19 Allan Variance of the gyroscope's zero rate output, showing ARW and an in-run bias instability for FRB and whole-angle operation.

and FTR loops is shown in Figure 3.4 [22]. A frequency mismatch of $\Delta f < 0.1$ Hz was observed along the x and y geometric axes, and a higher PLL frequency up to 0.5 Hz was observed at a 45° pattern angle. A nominal drive voltage of 1 mV was used for amplitude control. The maximum variation in quadrature null command signal was <25 mV, the variation on FTR command signal was <5 mV. Uncompensated quadrature null and FTR loops showed 2θ and 4θ dependence on pattern angle, Figure 5.20.

In order to operate the gyroscope in the whole-angle mode and permit precession of the vibration pattern, the FTR loop was disabled. The RIG operation was demonstrated by applying a continuous rate input of 180 °/s over 2 h duration on

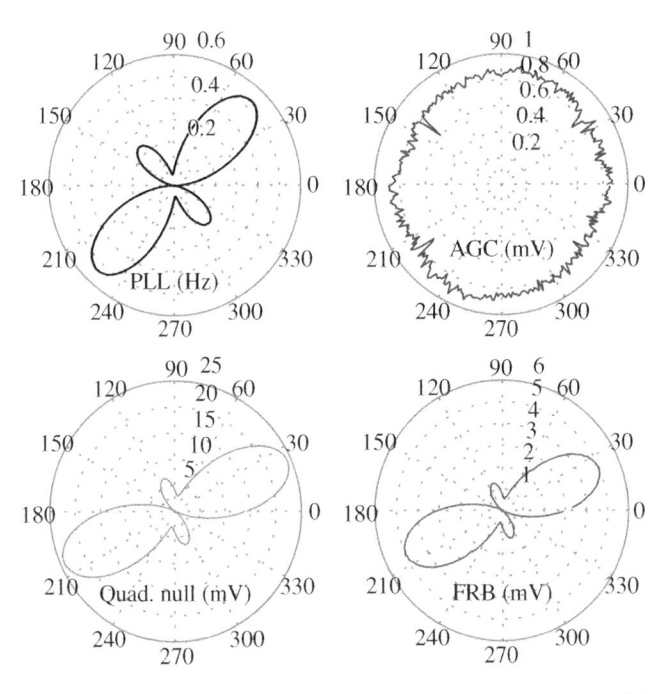

Figure 5.20 Polar plots showing the pattern angle dependence of four main closed-loops. 2Θ and 4Θ dependence indicate frequency mismatch and forcer misalignment.

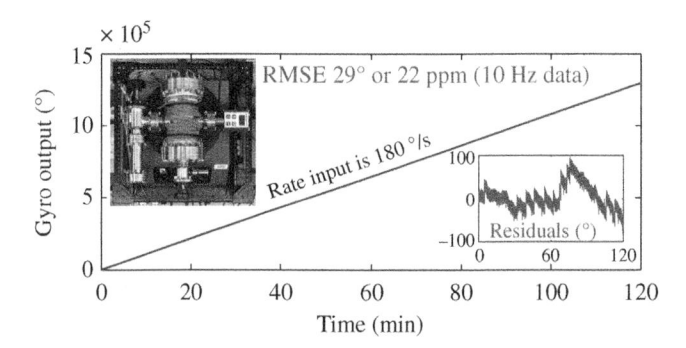

5.2 A Lumped Mass MEMS Gyroscope – Dual Foucault Pendulum Gyroscope

Figure 5.21 Spooling of the whole-angle gyro output over 2 h of continuous rotation. Linear fit shows angular gain factor of ~0.8 and RMS error of 22 ppm.

a rate table, Figure 5.21. A linear fit to the experimental data revealed an angular gain factor of ~0.8. Local perturbations of >5° were observed in the uncompensated gyro output due to a pattern angle dependent bias (angular error and pattern drift).

Part III

3-D Micro-Machined Whole-Angle Gyroscope Architectures

6

Overview of 3-D Shell Implementations

Motivated by the proven performance of macro-scale Hemispherical Resonator Gyroscopes (HRGs) [17], there has been a growing interest in 3-D MEMS micro-shell resonators for use in timing and inertial sensing applications [80]. Micro-shell architectures may enable a new class of high performance MEMS gyroscopes due to potential advantages in symmetry, low energy dissipation, and immunity to external vibrations.

Aside from the challenges associated with obtaining a high-Q resonator with low frequency split (Δf), defining electrodes on these 3-D MEMS structures with sufficiently small gaps and uniformity provides an additional challenge, factors such as alignment errors between the shell and the electrodes, cross-talk between electrodes, relatively large gaps created by assembly based techniques, and lack of scalability pose a challenge.

This chapter reviews the recent advances in 3-D shell micro-technology. Starting with a brief history of macro-scale shell resonator gyroscopes, this chapter reports novel fabrication processes and architectures for implementation of whole-angle gyroscopes at micro-scale. Majority of the focus is directed toward development of a surface tension and pressure driven micro-glassblowing paradigm, which is envisioned to serve as an enabling mechanism for wafer-scale fabrication of 3-D micro-wineglass gyroscopes.

6.1 Macro-scale Hemispherical Resonator Gyroscopes

Macro-scale fused silica HRG motivates the investigation of 3-D micro-wineglass structures for use as vibratory elements in MEMS applications. Current state-of-the-art HRGs are fabricated through macro-scale precision machining techniques, which lead to very small relative tolerances (defect to smallest feature ratio). The combination of low internal loss fused silica and high structural symmetry of precision machining processes can lead to very high Q-factors, in excess of 25 million [17].

Whole-Angle MEMS Gyroscopes: Challenges and Opportunities,
First Edition. Doruk Senkal and Andrei M. Shkel.
© 2020 The Institute of Electrical and Electronics Engineers, Inc. Published 2020 by John Wiley & Sons, Inc.

6 Overview of 3-D Shell Implementations

Figure 6.1 Northrop Grumman HRG uses a double-stemmed fused silica wineglass resonator at its core [17]. © AAS/AIAA.

Figure 6.2 SAGEM HRG uses a mushroom/bell type fused silica resonator [88]. © ESA.

The highest performance macro-scale Coriolis Vibratory Gyroscope (CVG) is the HRG [17] that has been under development since 1975 by various organizations such as Draper, Delco, Litton, and the current manufacturer Northrop Grumman, Figure 6.1. The gyroscope went through many design iterations over the years [81–83]; however, the extremely high performance of the HRG has always been associated with the precision machined fused silica "wineglass resonator," that constitutes the heart of the gyroscope. Due to the extremely high Q-factors, over 25 million, provided by the fused silica resonator, HRG demonstrates exceptional bias stability (0.000080°/h) and noise performance (ARW $0.000010°/\sqrt{h}$). However, the gyroscope is primarily used for aerospace applications due to its extremely high cost and relatively large size >30 mm. Despite the current development efforts by Northrop Grumman Corporation to further reduce the size of the gyroscope [84, 85], production complexity, cost, and size remains a prohibiting factor for large-scale integration.

Another HRG, REGYS 20, has been developed and commercialized by French SAGEM [86, 87] to provide a reliable sensor for a medium performance class IMU [86], Figure 6.2. The gyroscope employs a precision machined fused silica resonator at its core, but with a mushroom/bell type resonator with an internal stem structure. One of the key advantages of such a design is the low number of moving parts (less than 5), compared to more than 30 parts for a ring laser gyro (RLG) or fiber optic gyro (FOG). The possibility of integrating planar electrode structures was also proposed as part of this design, which can help lower the fabrication cost associated with outer electrode fabrication [89–92].

6.2 3-D Micro-Shell Fabrication Processes

3-D micro-shell resonator geometries can be divided into five main categories based on their cross-section profiles, Figure 6.3: (a) wineglass shells, (b) cylindrical shells, (c) spherical shells, (d) mushroom/bell type shells, and (e) bundt/birdbath shells. In this section, recent efforts in fabricating 3-D micro-shell resonators are discussed. The section is divided into two main approaches: (i) micro-shell resonators fabricated through plastic deformation of bulk materials and (ii) micro-shell resonators fabricated through deposition of thin films onto predefined molds, Figure 6.4.

6.2.1 Bulk Micro-Machining Processes

In this section, we look at micro-shell fabrication processes that rely on plastic deformation of bulk materials, Figure 6.4. A process based on ultrasonic machining (and EDM) is also reviewed in this section.

Micro-glassblowing at wafer level was first demonstrated [93, 94] on borosilicate glass for fabrication of spherical cells to be used in chip-scale atomic devices [95]. The original plan for process development was to use conventional glassblowing techniques [93]. To accomplish this, borosilicate glass wafers were bonded to silicon wafers with through-etched holes and a blow-hose was used with the goal of blowing micro-spheres when heat is applied. However, forming an adequate seal between the blow-hose and the wafer stack at the high temperatures

(a) Wineglass (b) Cylindrical (c) Spherical (d) Mushroom/bell (e) Bundt/birdbath

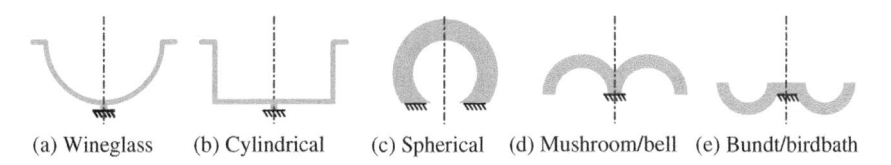

Figure 6.3 Cross-sectional view of various micro-shell resonator geometries: (a) wineglass shells, (b) cylindrical shells, (c) spherical shells, (d) mushroom/bell type shells, and (e) bundt/birdbath shells.

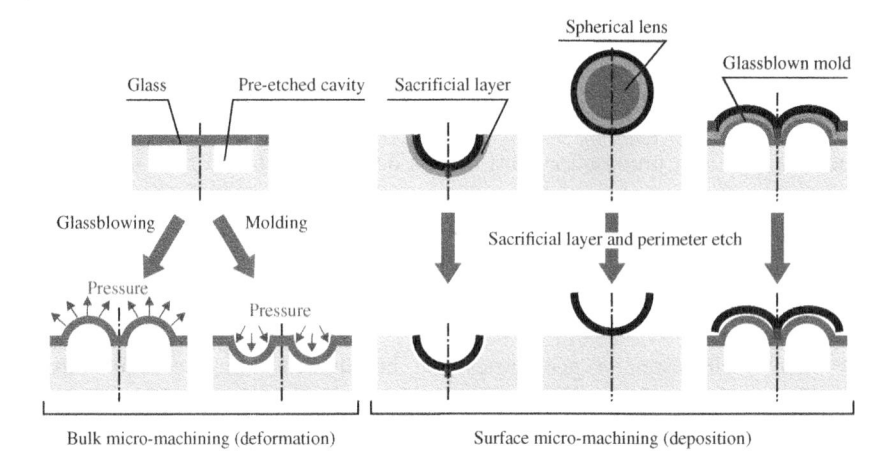

Figure 6.4 Micro-shell fabrication processes can be categorized into two main categories: (i) micro-shell resonators fabricated through plastic deformation of bulk materials, and (ii) micro-shell resonators fabricated through deposition of thin films onto predefined molds.

required for glassblowing proved to be difficult. While pursuing this approach, small air bubbles were observed at other locations on the surface of the wafer, which was hypothesized to be forming due to air pockets trapped between the glass and silicon wafers. It was later found that a similar process has been used to create micro-structures out of silicon [96].

Motivated by these initial results, a process was developed around the idea of amplifying the effects of air pockets. The so-called micro-glassblowing process utilizes an etched cavity on a silicon substrate wafer and a borosilicate glass layer that is bonded on top of this cavity, creating a volume of trapped gas for subsequent glassblowing of self-inflating spherical shells. When the bonded wafer stack is heated above the softening point of the borosilicate glass layer (around 850°C), two effects happen simultaneously: (i) the borosilicate glass layer becomes viscous, and (ii) the air pressure inside the pre-etched cavity increases above the atmospheric level. This results in plastic deformation of the glass layer, driven by gas pressure and surface tension forces (glassblowing). The expansion of air (and hence the formation of the shell) stops when the pressure inside and outside of the glass shell reaches an equilibrium, creating a self-limiting process. During this deformation, the surface tension acting on the now viscous glass layer works toward minimizing the surface area of the structure, and as a result a highly symmetric spherical shell with low surface roughness forms. The process allows simultaneous fabrication of an array of identical (or different, if desired) shell structures on the same substrate, Figure 6.5.

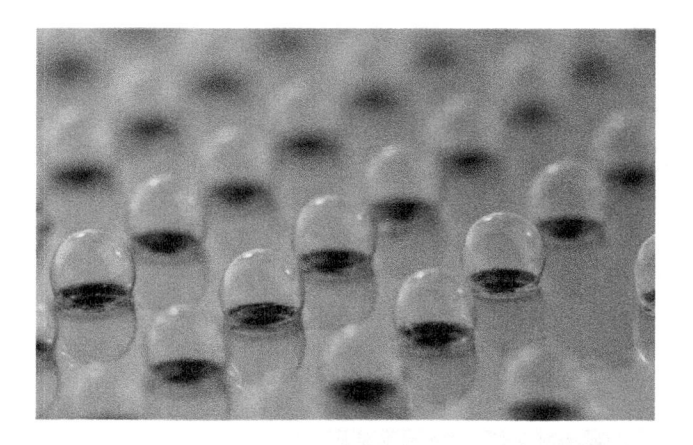

Figure 6.5 Arrays of spherical shells were created by bonding borosilicate glass wafer to a pre-etched silicon wafer and heating the wafer stack to 850°C [93]. © IEEE.

To enable electrostatic or magnetic transduction, a metal deposition step was later added to the micro-glassblowing process. This additional process step allows fabrication of 3-D metal traces on the surface of the glass shells [97, 98]. This is accomplished by: (i) patterning 2-D metal traces onto the glass surface while the wafer stack is flat, (ii) heating the wafer stack, which causes the glass layer to plastically deform and form 3-D metal traces on the surface of the spherical shells, Figure 6.6.

Another micro-shell fabrication process was developed at Yale University, based on plastic deformation of Bulk metallic glass (BMG) [99, 100], Figure 6.7. In this work, hemispherical shells were fabricated by blow molding Platinum-based ($Pt_{57.5}Cu_{14.7}Ni_{5.3}P_{22.5}$) BMG at a temperature of 275°C. Inert gases were used during most of the processing steps, due to low oxidation stability. Primary advantages of BMGs are low processing temperatures compared to most glass

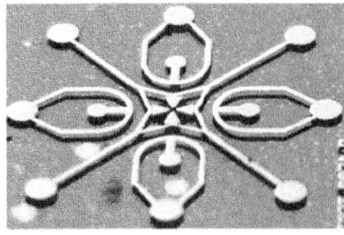

(a) Patterned 2-D metal traces (b) 3-D metal traces on shell surface

Figure 6.6 3-D metal traces can be fabricated on the surface of glass shells by (a) patterning 2-D metal traces on the wafer stack, (b) micro-glassblowing to plastically deform the metal traces into a 3-D geometry [97, 98]. © IEEE.

Figure 6.7 Bulk metallic glass shell structures are inherently conductive, eliminating the need for the additional metallization step needed for fused silica shell structures [99]. © IEEE.

materials, as well as inherent conductivity of the material, eliminating the need for metallization needed for fused silica shell structures [101].

Another micro-shell fabrication technique based on plastic deformation of glass was developed in [102] to create birdbath (hemitoroidal) and hemispherical shell structures, Figure 6.8. The primary difference from micro-glassblowing

Figure 6.8 Blow-torch molded birdbath shell resonator [103].

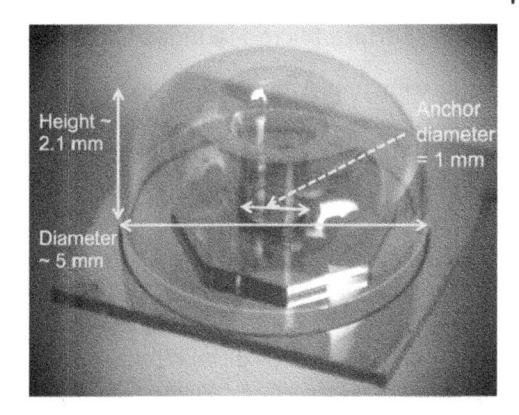

Figure 6.9 Fused silica spheres were micro-machined into 3-D shell structures using a combination of ultrasonic machining (USM), electro-discharge machining (EDM), and lapping [105]. © IEEE.

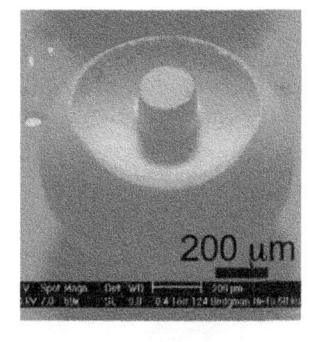

process [13, 93, 94] is the use of a blow torch to provide localized heat above 2500°C, as opposed to inserting the wafer stack into a furnace. Instead, to create the fused silica shells, thin layers of fused silica pieces were individually pressed onto graphite fixtures and deformed one at a time using the heat from a blow torch. Shell structures were later lapped from the backside to release the devices around their perimeter. Finally, the shells were sputter coated with a thin layer of Ti/Au for conductivity. Later a micro-welding step was added to the process to create micro-wineglass structures with arbitrarily sized fused silica stems [104]. This is achieved by inserting a fused-silica rod into the graphite fixture, which results in the fused silica device layer to flow and weld onto the rod.

Finally, a micro-machining process that utilizes ultrasonic machining (USM), electro-discharge machining (EDM), and lapping was proposed (3D-SOULE) to create micro-wineglass structures [105], Figure 6.9. EDM was mainly used to shape the stainless steel tooling, which was then used to USM fused silica spheres. Fused silica spherical-concave and mushroom type structures were created using this process. Laser Doppler Vibrometry was used to characterize the micro-wineglass structures, showing a Q-factor of 345 at 1.38 MHz in air.

6.2.2 Surface-Micro-Machined Micro-Shell Resonators

Surface-micro-machined micro-shells are almost exclusively fabricated by depositing a thin film onto a predefined mold with a sacrificial layer to create the resonator element, Figure 6.4. This is typically followed by a CMP step or perimeter etch to release the micro-shell structure. Due to the nature of thin film deposition processes, micro-shells of this type exhibit small size (<1–2 mm diameter) and thin structures (<5 µm thickness).

Isotropic wet etching of silicon molds using HF-HNO_3 and silicon nitride molds have been investigated at Cornell University [106], with the goal of depositing a thin film material (i.e. silicon nitride) into the mold at a later step to create hemispherical shell structures, Figure 6.10. Authors experimented with different HF and HNO_3 ratios as well as different silicon orientations, < 100 > and < 111 > wafers. The mold isotropy was analyzed using optical profilometry. The level of anisotropy was measured using optical profilometry. Due to the crystalline nature of silicon, the hemispherical molds were deformed toward a square shape for < 100 > silicon and toward a hexagonal shape for < 111 > silicon. Lowest measured anisotropy of 1.4% was obtained for < 111 > silicon wafers using higher HF: HNO_3

Figure 6.10 Silicon dioxide shells were formed by isotropic etching of silicon using HNA, growing thermal oxide, releasing using XeF_2, and finally reflowing the oxide using CO_2 laser [107].

Figure 6.11 Hemispherical shell structures were fabricated by isotropically etching silicon cavities, thermally growing SiO_2 inside, and later removing the silicon mold using XeF_2 etching [108].

ratios. The process was later used to fabricate opto-mechanical light transducers, by growing thermal oxide in pre-etched cavity, releasing the shell structure using XeF_2 and finally re-flowing the thermal oxide layer using CO_2 laser [107], Figure 6.10.

Hemispherical shell structures were fabricated in [108] by isotropically etching silicon cavities, thermally growing SiO_2 inside and later removing the silicon mold using XeF_2 etching, Figure 6.11 [108]. As opposed to wet etching performed in [106], the molds in this work were created using a dry etching process (SF_6 plasma etching). A radial deviation of 3.37 µm along the perimeter at a diameter of 1105 µm was reported. Hemispherical shell structures were subsequently coated with TiN using atomic layer deposition (ALD) for electrical conductivity [109]. The process was later adapted to fabricate low coefficient of thermal expansion (CTE) INVAR-36 shell structures, by sputtering INVAR-36 alloy onto the pre-etched cavity, CMP of INVAR top surface to release the perimeter and etching away silicon underneath [110]. Another adaptation of the process was used to create Polysilicon shell structures [111] by subsequent deposition of sacrificial and PolySi layers into the pre-etched cavity.

Polycrystalline diamond hemispherical shell structures were fabricated in [112] by depositing polydiamond thin films into hemispherical molds on a silicon wafer, Figure 6.12. Primary advantages of polycrystalline diamond films are: potentially

Figure 6.12 SEM image of arrays of 1 mm diameter released polycrystalline diamond shells [112]. © IEEE.

high Q-factor due to low thermoelastic damping [114] and a potential for boron doping, creating inherently conductive shell structures, bypassing the need for an additional metal layer. Instead of wet/dry etching as in [106, 108], the hemispherical molds were created by μ-EDM (electro-discharge machining), followed by Au/Cr mask deposition and HNA (HF, nitric acid, acetic acid) wet etching to smoothen the mold surface, Figure 6.13 [113]. Afterward, a 2 μm thick SiO_2 sacrificial layer was deposited, followed by 1 μm boron doped hot filament CVD diamond deposition. The process was completed by doing back-side etch to define a Si_3N_4 stem structure, diamond dry etching along the perimeter and finally HF release. In another implementation of the process, 1500 and 625 μm deep polycrystalline diamond cylindrical shells were created by DRIE of cylindrical cavities, which were subsequently annealead at 700°C for 2 h to improve the surface roughness [115].

All-dielectric (SiO_2) cylindrical gyroscopes were reported by HRL Laboratories [116, 117], Figure 6.14. The main difference from cylindrical resonators in [118] is the SiO_2 resonator material.

Another polydiamond HRG was reported by Charles Stark Draper Laboratory [119], Figure 6.15. In this research, wet etching of Corning 1715 glass was used to achieve highly isotropic cavities compatible with temperatures required for polydiamond deposition, while retaining a closer coefficient of expansion match (CTE) to the polycrystalline diamond structure. Using this technique, average cavity diameters of 1288 μm were etched, with perfect roundness within the resolution of the measurement (±0.5 μm).

Another SiO_2 hemispherical shell fabrication process was reported by University of Utah in [120]; this process also relies on isotropically etched hemispherical

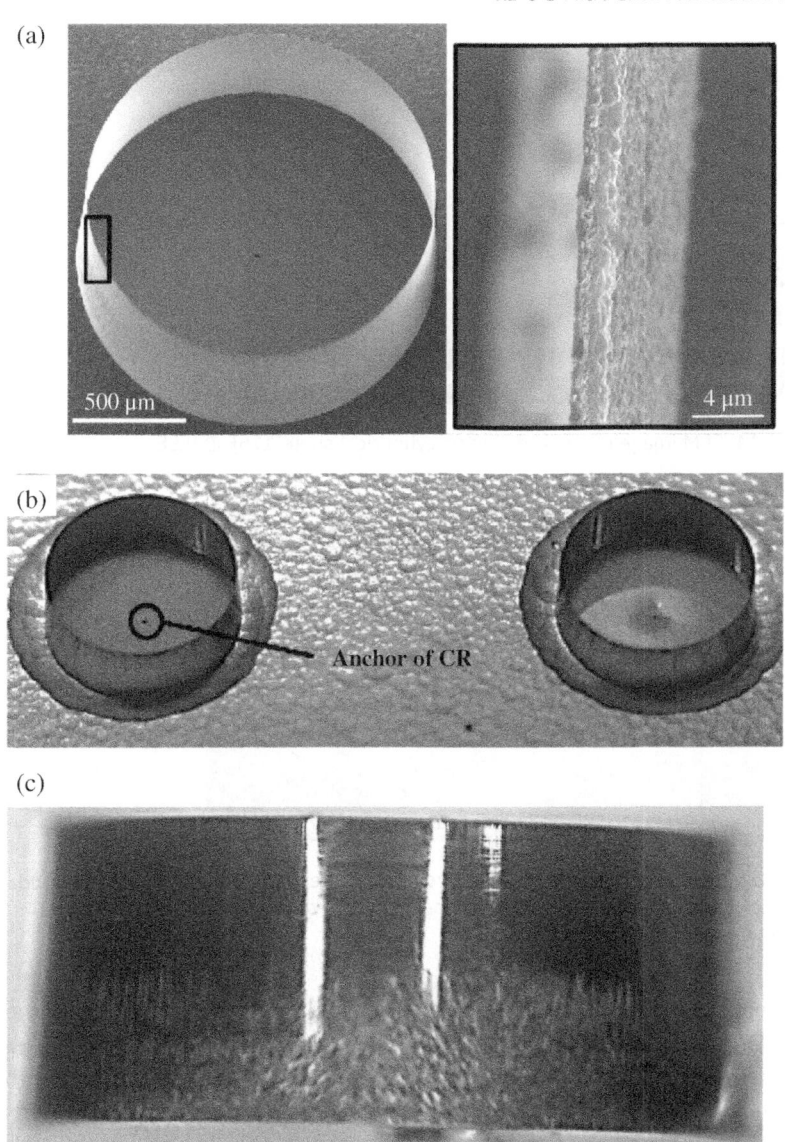

Figure 6.13 Cylindrical polycrystalline diamond shells can be created if the hemispherical etch is replaced with DRIE of cylindrical cavities [113]. (a) Image showing shell perimeter, (b) top-down view of two cylindrical shells after XeF_2 etch, (c) side view of the shell showing roughness due to DRIE etch of the mold [115]. © IEEE.

Figure 6.14 SEM image of an all-dielectric cylindrical shell [116]. © IEEE.

Figure 6.15 Diamond hemisphere deposited into a pre-etched glass cavity and released. Process is enabled by extremely high etch selectivity of diamond to all known room temperature etches [119]. © IOP.

molds on a silicon wafer. Thermally grown SiO_2 was used as an etch stop layer along with a polysilicon sacrificial layer underneath the oxide shells. Piezo-actuation and electrostatic drive using a probe tip were used for testing with laser Doppler vibrometry pick-off. Later electrostatic transduction and Q-factors above 10 000 at 22 kHz center frequency were reported [121]. In addition, thin film sputtered ULE (Ultra Low Expansion Glass) shells were reported using a process called "Poached-Egg Micro-molding" [122]. As opposed to using hemispherical molds on a silicon wafers, the authors utilized precision ball lenses as a mold. The ball lenses were coated with a polysilicon sacrificial layer followed by sputtering of ULE glass as the device layer. The coated ball lenses were placed onto silicon posts and the ULE above the equator line of the lens was etched using Ar plasma etching. Subsequently the ball lens was removed by etching the ULE above

Figure 6.16 Thin film sputtered ULE (Ultra Low Expansion Glass) shells were reported using a process called "Poached-Egg Micro-molding," which utilizes an off-the-shelf ball lens as the mold [120]. © IEEE.

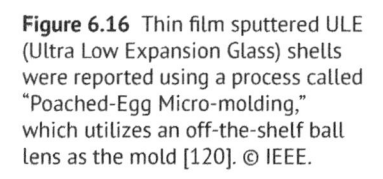

the equator of the ball lens and XeF_2 of the polysilicon device layer, leaving a sputtered ULE shell structure in the shape of the ball lens, Figure 6.16.

Polycrystalline diamond half-toroidal shells were reported by Honeywell International in [123], Figure 6.17. Micro-shells were fabricated by depositing 0.8–1.5 µm polycrystalline diamond onto micro-glassblown hemitoroidal molds along with a polysilicon sacrificial layer. Deposition was done by sp3 Diamond technologies using a hot-filament CVD process.

Extremely small (200 µm diameter) cenosphere-derived hemispherical shells were reported by University of Michigan in [124], Figure 6.18. The shells are fabricated from sodium borosilicate cenospheres that are commercially available from Potters Industries LLC. The shells are fabricated by expansion of gases within the glass particles, which act as blowing agents. The hemispherical shells are created by attaching borosilicate glass cenospheres to a silicon substrate and ion-milling a 400 nm pattern at 10 kV and 26 nA in order to keep the shells from breaking.

6.3 Transduction of 3-D Micro-Shell Resonators

In order to achieve functional resonators based on micro-shell structures, efficient methods for transduction are required. Unlike conventionally micro-machined 2-D MEMS resonators, 3-D geometry of micro-shell structures and high aspect ratios pose a significant challenge for efficient transduction. In this section, various transduction schemes for micro-shell resonators are reviewed.

6.3.1 Electromagnetic Excitation

In [97, 98], 3-D integrated electromagnetic transducers were fabricated on a micro-glassblown shell resonator by depositing flat metal traces on borosilicate

Figure 6.17 Hemitoroidal polycrystalline diamond shell structure [123]. © IEEE.

glass surface and subsequently micro-glassblowing, Figure 6.6. A Cr/Cu/Au metal stack with 0.8 μm Cu was found optimal, where the Cr layer provides adhesion, the ductile Cu layer prevents segmentation, and Au layer prevents oxidation of the metal stack. For excitation, a 0.2 mA current was applied to a metal trace in the presences of 0.5 Tesla magnetic field, provided by a permanent magnet. LDV detection revealed a 687 kHz resonance peak corresponding to rocking mode of the micro-shell resonator.

6.3.2 Optomechanical Detection

In [107], opto-mechanical detection was demonstrated on a SiO_2 micro-wineglass. CO_2 laser was used to reflow the perimeter of the micro-wineglass to form an optical cavity along the perimeter of the micro-wineglass, Figure 6.10. A tapered fiber was used to couple light into the optical cavity of the micro-wineglass. Later, mechanical modes of the micro-wineglass were excited using piezo actuation. The mechanical motion resulted in modulation of light intensity inside the tapered fiber, which was detected by an Agilent

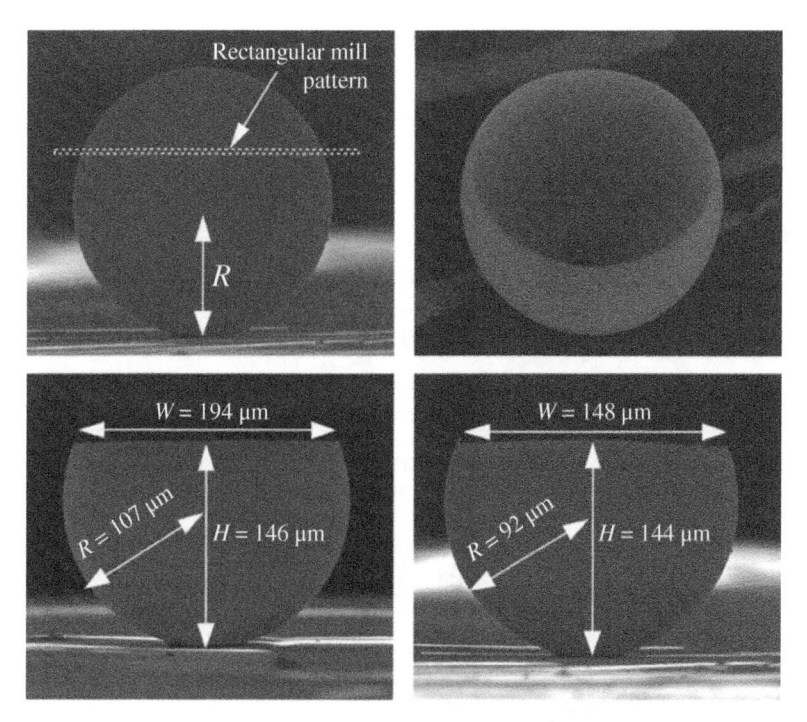

Figure 6.18 SEM image of extremely small (200 μm diameter) cenosphere-derived hemispherical shells [124]. © IOP.

N5320 Network Analyzer. $n = 2$ and $n = 3$ wineglass modes were successfully characterized using this method, demonstrating the viability of micro-wineglass structure for use in opto-mechanical gyroscopes.

6.3.3 Electrostatic Transduction

An assembled electrode structure for initial characterization of early prototypes was demonstrated in [125]. Thru-holes were etched on substrate wafer of the micro-shell structure. On a separate wafer, Silicon-On-Insulator (SOI) electrode pillars were fabricated using DRIE. Electrodes were assembled by inserting the electro pillars through the thru-holes on the micro-shell substrate wafer.

In [116, 117], electrostatic actuation of micro-shell resonators using electric field gradients was demonstrated. Initial demonstration was done using an assembled interdigitated metal electrode pattern placed in close proximity of the micro-shell resonator [117]. Differential voltage applied to adjacent electrodes generates fringe fields which interact with the micro-shell structure. In [117], metal traces were replaced with interdigitated silicon electrodes, Figure 6.19. In both instances, the

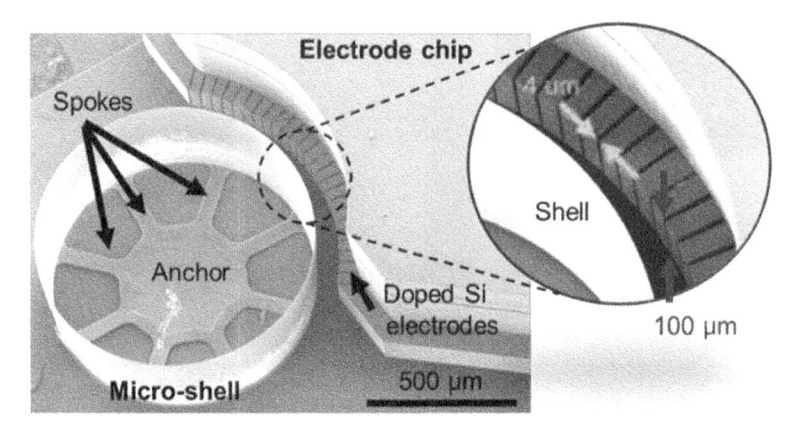

Figure 6.19 Highly doped silicon electrodes adjacent to a SiO_2 micro-shell, fringe fields between interdigitated fingers excite the micro-shell structure [117].

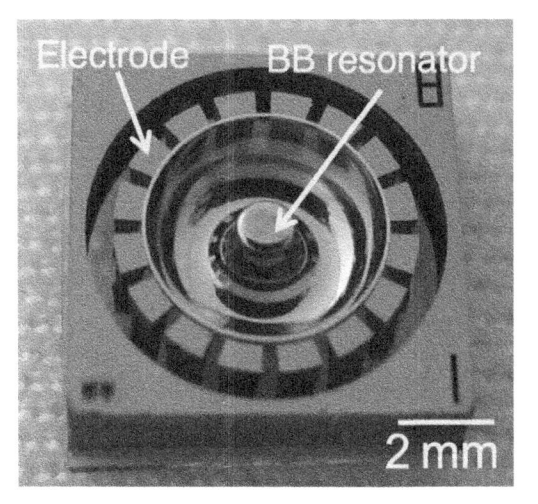

Figure 6.20 Blow-torch molded fused silica micro-shell resonator with silicon electrodes, diameter 5.0 mm, height 1.55 mm [126]. © IEEE.

detection was done using LDV measurement. This approach bypasses the need for metallization of the micro-shell structure for transduction.

In [126], silicon electrodes were assembled on a fused silica blow-torch molded micro-shell resonator. The micro-shell resonator was fabricated using the blow-torch molding process [103]. Electrodes were fabricated by: (i) etching 1.6 mm deep trenches on a P-type 2 mm thick silicon wafer, (ii) evaporating a 4 μm Al protection layer, (iii) bonding to a borosilicate glass substrate, (iv) removing the remaining thickness of the silicon, and (v) blanket metallization of the micro-shell resonator, followed by assembly onto the electrode structure, Figure 6.20. Across the 16 electrodes, gaps in the range of 10.3–16.3 μm were measured, averaging around 14.2 μm.

6.3 Transduction of 3-D Micro-Shell Resonators | 83

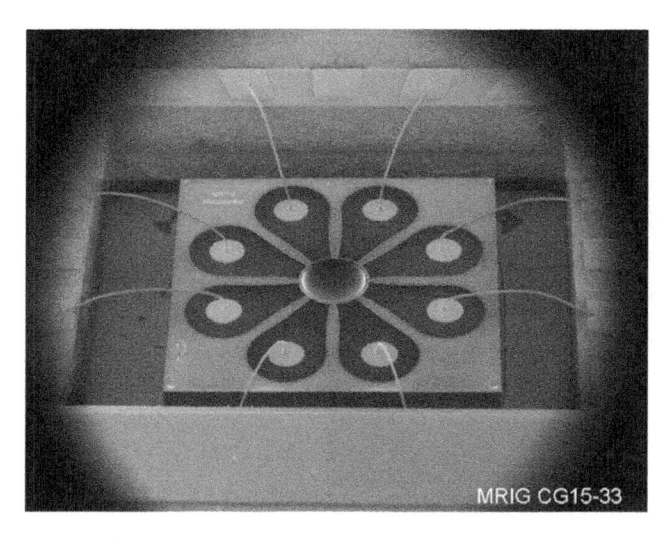

Figure 6.21 Polydiamond micro-shell resonator with integrated electrodes [119]. © IOP.

Polycrystalline diamond micro-shell resonators with co-fabricated electrodes were fabricated in [119], Figure 6.21. Corning 1714 glass was used as a substrate for its high softening point. The substrate was wet-etched and coated/patterned with 2 µm thick sacrificial polysilicon, followed by 2 µm thick boron-doped polydiamond deposition, which was used for fabrication of both the micro-shell resonator and the electrodes. After polydiamond fabrication, Cr/Au bond pads were deposited onto the electrodes and sacrificial polysilicon layer was removed using XeF_2. Mode-shapes and frequencies of the resonator were characterized using piezo or electrostatic actuation and LDV detection. A similar fabrication flow for polydiamond wineglass resonators was developed in [123], Figure 6.17. Primary difference to the process demonstrated in [119] is the use of convex micro-glassblown structure instead of a concave wet-etched cavity for the mold.

A variation of this process using a silicon substrate was developed in [127]. Surface of the silicon substrate was coated with 1.2 µm LPCVD SiN layer, which was subsequently coated and patterned with doped 0.5 µm polysilicon and 1 µm sacrificial SiO_2 layers. Finally the polydiamond layer was coated and patterned. To release the resonator, SiO_2 layer was removed using HF etching, forming capacitive gaps between the doped polysilicon electrodes and the polydiamond resonator. Fully electrostatic transduction was demonstrated using this process.

Platinum-based Bulk Metallic Glass (Pt-BMG) micro-shell resonators with integrated electrodes was demonstrated in [99]. The electrodes were defined on the surface of a pre-etched glass mold. A thru-hole was created at the center of the mold cavity using laser ablation. A steel pin was inserted through the thru-hole,

Figure 6.22 Bulk Metallic Glass (BMG) micro-shell resonator with integrated electrodes. Thermal expansion mismatch between the micro-shell and the mold helps to remove the micro-shell [99]. © IEEE.

Figure 6.23 Micro-shell resonator with polysilicon electrodes [111]. The electrodes were deposited into DRIE trenches in the silicon mold.

which would later form the stem of the micro-shell resonator. Later Pt-BMP material was molded onto the pre-etched cavity (and the pin) at a temperature of 275°C and 0.69 MPa pressure. The temperature mismatch between the mold and the glass mold allowed separation of the micro-shell from the mold once cooled, creating capacitive gaps, Figure 6.22.

In [128], silicon with different doping concentrations was used to define the electrodes on the micro-shell substrate, and electrodes were isolated from the substrate (and each other) using PN junctions. A further extension of this approach, DRIE trenches were etched into the micro-shell substrate, which were subsequently coated with SiN for isolation and back-filled with polysilicon to form the discrete electrodes [111], Figure 6.23. A capacitive gap of 20 μm was obtained using this process.

Figure 6.24 Silicon electrodes for "Poached Egg" micro-shell resonators [118]. © IEEE.

Silicon electrodes for "Poached Egg" micro-shell resonators was demonstrated in [118], Figure 6.24. A thick silicon on glass wafer was etched with DRIE to create high aspect ratio silicon electrodes and coated with Parylene for protection. To create the micro-shell structure, a 1 mm spherical lens was used as the mold and coated with 3 µm polysilicon sacrificial layer, followed by 1.2 µm sputtered Ultra Low Expansion (ULE) glass. Later, the spherical ball bearing was bonded into the electrode housing. Top half of the spherical lens was etched using anisotropic etching and the spherical lens was carefully removed, revealing the micro-shell resonator in the center of the electrode assembly.

7

Design and Fabrication of Micro-glassblown Wineglass Resonators

This chapter discusses an intriguing possibility: micro-glassblowing as a means to fabricate low internal loss fused silica and Ultra Low Expansion Titania Silicate Glass (ULE TSG) 3-D shell micro-structures. In Section 7.1, the design parameters for micro-glassblown structures is presented. This is followed by details of the fabrication process in Section 7.2.

As described in Chapter 2, maximization of the Q-factor is key to enhancing performance of vibratory MEMS devices in demanding signal processing, timing, and inertial applications [5]. Current MEMS fabrication techniques limit the maximum achievable Q-factor by restricting the material choice to few materials and device geometry to 2-D planar structures. Available materials such as single-crystal silicon have relatively high thermoelastic dissipation and 2-D planar devices are mostly limited by anchor losses. The macro-scale Hemispherical Resonator Gyroscope (HRG) with Q-factors over 25 million [17] motivates the investigation of 3-D fused silica micro-wineglass structures for use as vibratory elements in MEMS applications.

It has been demonstrated that single-crystal silicon MEMS devices can reach the fundamental Q_{TED} limit by using a combination of balanced mechanical design and vacuum packaging with getters [129]. TED is caused by local temperature fluctuations due to vibration and the associated irreversible heat flow, which results in entropic dissipation. TED can be reduced either by decoupling the frequencies of mechanical vibrations from the thermal fluctuations or by using materials with low CTE. This chapter focuses on materials with low CTE, such as fused silica (0.5 ppm/°C) and ultra low expansion TSG (0.03 ppm/°C), which can provide a dramatic increase in fundamental Q_{TED} limit ($Q_{TED} > 7 \times 10^{10}$ for a TSG wineglass structure). When compared to silicon, TSG and fused silica dry etching suffers from an order of magnitude higher surface roughness, lower mask selectivity, ~1 : 1 for KMPR photoresist and lower aspect ratio, <5 : 1 [20, 21].

Pyrex glassblowing at 850°C on a silicon substrate was previously demonstrated for fabrication of smooth, symmetric 3-D structures [93, 97, 98, 130]. Fused silica/TSG glassblowing requires temperatures upward of 1600°C due to its higher softening point, which prevents the use of fabrication processes that rely on a silicon substrate. This chapter discusses how the high temperature glassblowing (up to 1800°C) can be used for wafer-scale fabrication of TSG and fused silica 3-D wineglass structures.

7.1 Design of Micro-Glassblown Wineglass Resonators

Micro-glassblowing process utilizes an etched cavity on a substrate wafer and a glass layer that is bonded on top of this cavity, creating a volume of trapped gas for subsequent glassblowing of self-inflating spherical shells. When the bonded wafer stack is heated above the softening point of the structural glass layer, two effects are activated at the same time: (i) the glass layer becomes viscous, and (ii) the air pressure inside the pre-etched cavity increases above the atmospheric level. This results in plastic deformation of the glass layer, driven by gas pressure and surface tension forces (glassblowing). The expansion of air (and hence the formation of the shell) stops when the pressure inside and outside of the glass shell reaches an equilibrium, creating a self-limiting process. During this deformation, the surface tension acting on the now viscous glass layer works toward minimizing the surface area of the structure as a result of a highly symmetric spherical shell with low surface roughness forms. The process allows simultaneous fabrication of an array of identical (or different, if desired) shell structures on the same substrate.

Shape and size of the final glassblown structure can be designed by changing starting conditions such as thickness of the device layer, cavity shape, and dimensions, as well as environment temperature and pressure during glassblowing [93]. For example, changing the cavity diameter directly affects the diameter of the final glassblown shell, whereas changing the cavity depth (or volume) affects the height of the glassblown structure. It is also possible to fabricate entirely different geometries by changing the initial conditions, for example a circular cavity creates a spherical shell, when glassblown, Figure 7.1 (right). Inverted-wineglass structures can also be fabricated by defining a central post inside the etched cavity, Figure 7.1 (left). When the device layer is bonded to this central post, it acts as an anchor point, which allows the glassblown shell to fold around it, creating a self-aligned stem structure as in Figure 7.1e.

Figure 7.1 ULE TSG/fused silica micro-glassblowing process, consists of: (a) Deep fused silica etching of device footprint into the substrate, (b) plasma-assisted bonding of TSG/fused silica device layer, (c) micro-glassblowing at temperatures up to 1700°C, and (d) optional release step. Fabricated TSG and fused silica structures are shown in (e) [13]. © IEEE.

(a) Footprint of the desired geometry is patterned on fused quartz.

(b) Fused quartz is etched and TSG device layer is bonded on top.

(c) TSG/fused quartz stack is glassblown at 1600–1700°C.

(d) Wineglass structure is released along the perimeter.

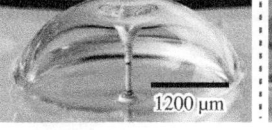

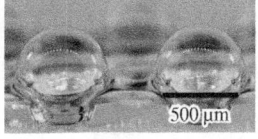

(e) Fabricated wineglass structure (left) and spherical shells (right).

7.1.1 Design of Micro-Wineglass Geometry

In this section, first analytical expressions to predict the final micro-glassblown geometry will be developed. This will be followed by finite element methods (FEMs) to predict the thickness of the shell structure and stem diameter.

In micro-glassblowing, the final device geometry heavily depends on the photolithographic pattern on the wafer surface as well as the etch depth of the cavity. For example, by using a device layer thickness of 200 µm, solid self-aligned stem structures were obtained for a central post diameter of 400 µm, whereas hollow hemitoroidal structures were obtained for a central post diameter of 600 µm, Figure 7.2. For this reason an accurate method to estimate the final geometry from initial dimensions is required.

7.1.1.1 Analytical Solution

In this section, analytical expressions for predicting the dimensions of the final inverted-wineglass structure are derived [131]. These expressions can be used to calculate the height and minor radius of the structure (h and r) based on the initial cavity dimensions. These expressions assume ideal hemitoroidal shell structures with zero thickness; as such it is not possible to predict the thickness of the wineglass shell or the diameter of the stem structure.

Calculation starts by finding the volume of the etched cavity:

$$V_{\text{cavity}} = \pi(r_2^2 - r_1^2)h_e, \tag{7.1}$$

where r_2 is the outer perimeter of the cavity, r_1 is the perimeter of the central post, and h_e is the etch depth, Figure 7.3. Upon heating, air inside the cavity will expand

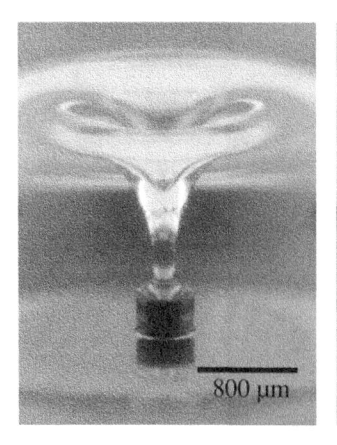

Figure 7.2 Small central post diameters create solid stem structures (left), large diameters create hemitoroidal structures (right).

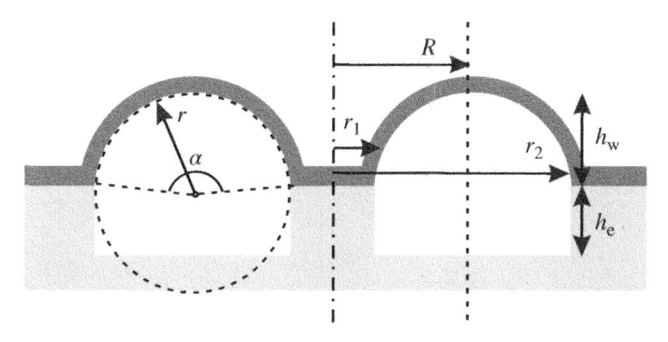

Figure 7.3 Geometric parameters of an inverted-wineglass structure: Minor radius r, major radius R, inner perimeter r_1, outer perimeter r_2, etch depth h_e, and wineglass height h_w.

to fill the volume of the wineglass shell. This volume can be calculated using the ideal gas law:

$$V_{\text{wineglass}} = \left(\frac{T_{\text{final}}}{T_{\text{initial}}} - 1\right) V_{\text{cavity}}, \tag{7.2}$$

where T_{initial} and T_{final} are the initial and final glass-blowing temperatures in degree kelvin. It is assumed that the air inside the cavity is at atmospheric pressure, which is also the pressure of the glassblowing chamber. The volume of the wineglass can also be calculated from geometric parameters using:

$$V_{\text{wineglass}} = \pi R r^2 (\alpha - \sin(\alpha)), \tag{7.3}$$

where r is the minor radius, R is the major radius of the partial toroid, and α is the fullness parameter in radians (central angle of the arc formed by the minor radius), Figure 7.3. Minor and major radii can be removed from the above expression using:

$$r = \frac{r_2 - r_1}{2\sin(\alpha/2)}, \quad R = \frac{r_1 + r_2}{2}. \tag{7.4}$$

Substituting (7.1) into (7.2) and (7.4) into (7.3) leaves α to be the only unknown variable, which can be solved numerically. Once α is known, all other parameters of the glassblown shell structure can be extracted using geometric relationships in Figure 7.3. For example, a relationship between minor radius (r), fullness parameter (α), and wineglass height (h_e) can be given as:

$$h_w = r(1 - \cos(\alpha/2)). \tag{7.5}$$

Solutions of these expressions for a large variety of micro-wineglass structures are presented in Figure 7.4. The expressions presented in this section are not sufficient to calculate the shell thickness or the stem diameter; FEMs to calculate these parameters will be presented in the next section.

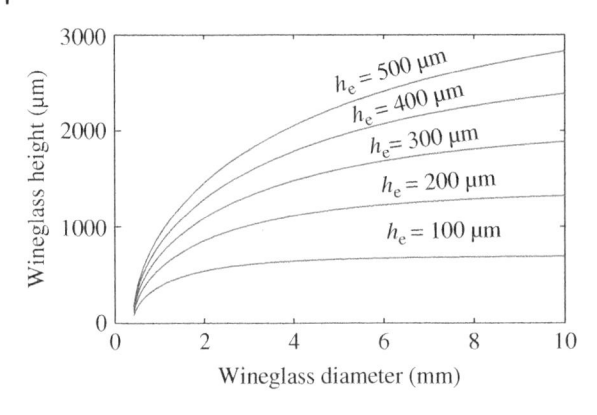

Figure 7.4 Analytical solution of etch depth (h_e), wineglass diameter ($2r_2$) versus final inverted wineglass height. Stem outer diameter ($2r_1$) is 400 μm, glassblowing temperature is 875°C.

7.1.1.2 Finite Element Analysis

Analytical expressions presented in (7.1) through (7.4) are not sufficient to predict the shell thickness and the stem diameter. For this reason, FEM models for the micro-glassblowing process were developed to predict the effect of subtle changes in initial dimensions on the final geometry [131].

Due to the large deformation of the shell structure, Arbitrary Lagrangian-Eulerian (ALE) technique [132, 133] was used. ALE allows the mesh to deform, as to track the deformation of the structure in the time domain and reapply the boundary conditions at every time step. Comsol Multiphysics Package was used for the analysis; the following assumptions were used for boundary conditions:

- At the glassblowing temperature (>850°C for borosilicate glass and >1600°C for fused silica), the deformation of the glass can be modeled using viscous fluid flow with a viscosity of 1×10^3 to 1×10^6 Pa.s [134].
- The driving force is a slowly varying (quasi-static) uniform pressure field within the shell cavity, Figure 7.5.
- Initial pressure inside the cavity is equal to atmospheric pressure.
- Outer surface of the shell is exposed to atmospheric pressure ($P_{gauge} = 0$).
- The surfaces that are bonded to the substrate are not moving (no-slip condition).
- The shells are axisymmetric as such a 2-D axisymmetric model with <1000 elements is sufficient for solution.

Using the above assumptions, the gauge pressure inside the cavity can be written as:

$$P_{internal} = \frac{T_{final} P_{initial} V_{cavity}}{T_{initial}(V_{cavity} + V_{wineglass})} - P_{initial}, \quad (7.6)$$

where T is the temperature in degree kelvin and $P_{internal}$ is applied uniformly to the inner surface of the micro-wineglass structure during glassblowing, Figure 7.5.

Figure 7.5 Boundary conditions for finite element analysis: (a) before glassblowing and (b) after glassblowing.

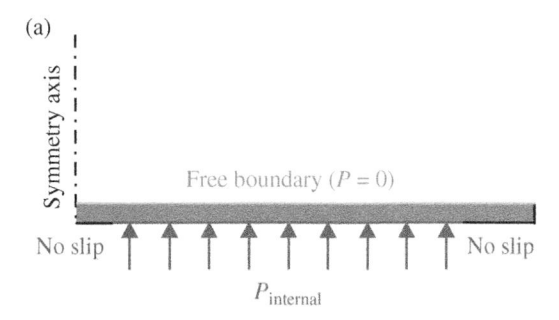

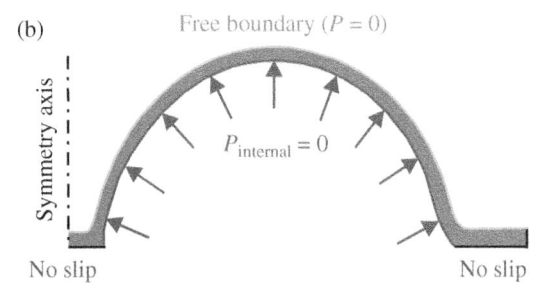

(a) Symmetry axis; Free boundary ($P = 0$); No slip; No slip; $P_{internal}$

(b) Symmetry axis; Free boundary ($P = 0$); $P_{internal} = 0$; No slip; No slip

Since the volume of the wineglass will continuously change during the transient solution, (7.3) cannot be used to calculate $V_{wineglass}$. Instead, a surface integral for the inner surface of the wineglass is used:

$$V_{wineglass} = \oint 2\pi r'^2 dnr, \qquad (7.7)$$

where r' is the distance of any point in the shell structure from the symmetry axis and dnr is the projection of the infinitesimal surface area onto the symmetry axis. Equation (7.7) allows continuous calculation of the shell volume and consequently the cavity pressure as the structure deforms. This allows the model to reach equilibrium when the final volume is reached ($P_{internal} = 0$).

Figure 7.6 shows the time domain solution of the micro-glassblowing process and the formation of the self-aligned stem structure. For a device layer thickness of 100 μm, decreasing the central post diameter from 400 to 200 μm is sufficient to change the shell structure from a hemitoroidal geometry to an inverted-wineglass with a solid stem structure. Figure 7.7 shows a side-by-side comparison of the finite element models and the actual fabricated geometries. The results from the models are compared to cross-sectional SEM shots in Table 7.1, showing ~20% accuracy for device 1 and better than 10% accuracy for device 2 in prediction of final geometry for device. This small variation is attributed to variation in furnace temperature from assumed values.

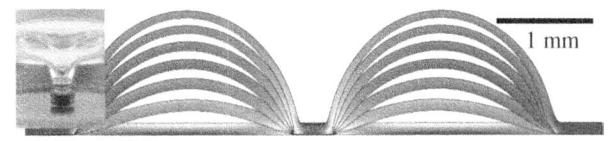

(a) 400 μm stem OD creates a hemitoroidal structure.

(b) 200 μm stem OD creates a mushroom structure.

Figure 7.6 Transient FEA of micro-glassblowing process showing the formation of self-aligned stem structures.

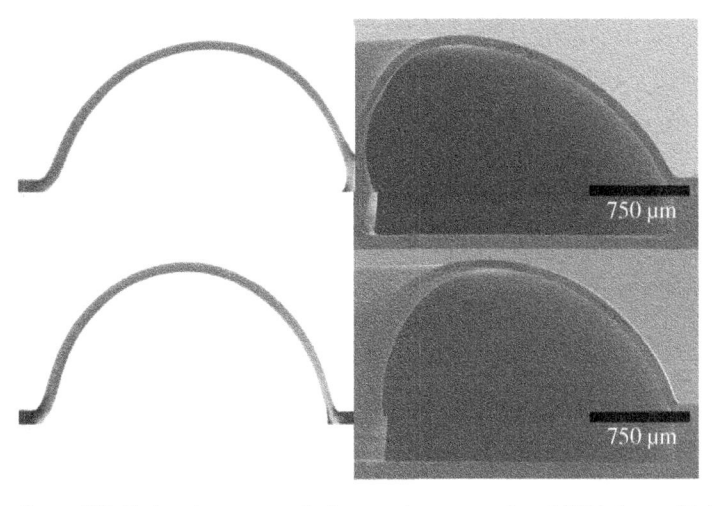

Figure 7.7 Finite element predictions and cross-sectional SEM shots of fabricated micro-wineglass structures [135]. © IEEE.

The idea of using Finite Element Analysis to design micro-glassblown wineglass resonators was further explored in [136] for the purpose of mode ordering. It was found that the $n = 2$ wineglass mode can be made to be the first vibration mode by controlling geometric parameters of the shell, such as anchor radius, as well as shell radius and thickness.

7.1.1.3 Effect of Stem Geometry on Anchor Loss

Anchor loss is the dissipation of kinetic energy within the vibrating structure into the substrate and the environment by means of acoustic waves, which ultimately

7.1 Design of Micro-Glassblown Wineglass Resonators

Table 7.1 Comparison of wineglass dimensions obtained from analytical solutions, FEA, and experiments.

			Device 1	Device 2
Initial thickness		(μm)	80	300
Outer diameter		(mm)	4.2	4.2
Etch depth		(μm)	240	200
Glassblowing temperature		(°C)	875	1700
Analytical	Final height	(μm)	971	1182
	Thickness	(μm)	N/A	N/A
FEA	Final height	(μm)	1025	1260
	Thickness	(μm)	52	205
Experimental	Final height	(μm)	832	1288
	Thickness	(μm)	44	229

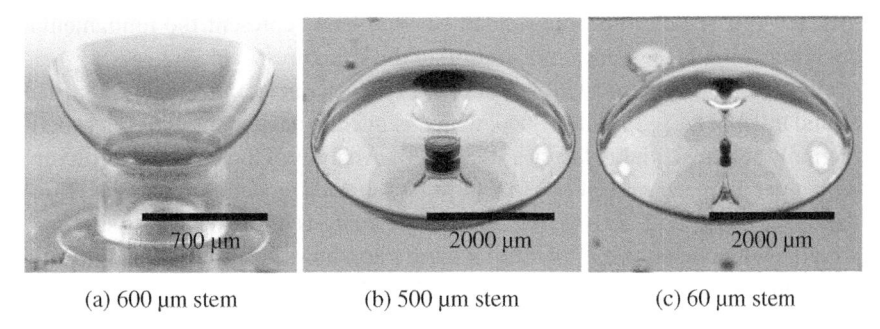

(a) 600 μm stem   (b) 500 μm stem   (c) 60 μm stem

Figure 7.8 Wineglass structures with (a) 1.2 mm outer diameter and 600 μm stem diameter (hollow), (b) 4200 μm outer diameter and 500 μm stem diameter (solid), and (c) 4200 μm outer diameter and 150 μm stem diameter (solid) [13]. © IEEE.

limits the overall Q-factor of the resonator. In this section, two types of geometries are analyzed to demonstrate the impact of stem diameter on Q_{anchor}: (i) Wineglass structures with hollow stem [130] shown in Figure 7.8a, and (ii) inverted wineglass structures as in Figure 7.8b and c. The structure in Figure 7.8a can be fabricated by first glassblowing a spherical structure through a stencil layer and then cutting the top half of the spherical shell using laser ablation to create a wineglass structure. Whereas the structures in Figure 7.8b and c can be fabricated by etching a toroidal cavity as described in Figure 7.1.

The structure in Figure 7.8a had a shell diameter of 1142 μm, anchor diameter of 600 μm, and average thickness of 4 μm, which gives roughly 1 : 2 attachment to

Table 7.2 Anchor loss analysis shows a large change in Q_{anchor} for different stem diameters.

Stem diameter (μm)	600	500	60
Wineglass diameter (μm)	1200	4200	4200
Q_{anchor}	1×10^3	2×10^6	5×10^{10}

shell diameter ratio. In contrast, the inverted-wineglass structures with the integrated stem, Figure 7.8b and c, had a shell diameter of 4200 μm, an average thickness of 80 μm with 500 μm and 150 μm stem diameters, respectively, giving a 1 : 8 and 1 : 28 stem to shell diameter ratios.

To simulate the acoustic loss in an infinite medium, a perfectly matched layer (PML) within Comsol Multiphysics Package can be used for modeling the substrate domain. PML works by absorbing acoustic waves over a large frequency range at any nonzero angle of incidence. The simulation was run for perfectly symmetric structures, neglecting the contribution of mass imbalance to the anchor loss. For this reason, the values obtained from FEA represent the fundamental anchor loss limit (theoretical maximum) of the structures, Table 7.2.

The wineglass structure with 1 : 2 anchor to shell diameter ratio, Figure 7.8a, had a fundamental Q_{anchor} limit of 3000, which is in close agreement with the experimentally obtained quality factor of 1256 [130]. Analysis of the wineglass structure with outer diameter ratio of 1 : 8 in Figure 7.8b showed moderate anchor loss (fundamental $Q_{anchor} > 2 \times 10^6$). In contrast, the analysis of the wineglass structure with the integrated stem (1 : 14 ratio), Figure 7.8b, shows potential for significantly lower anchor loss (fundamental $Q_{anchor} > 5 \times 10^{10}$).

7.1.2 Design for High Frequency Symmetry

Lumped mass gyroscopes, such as the Foucault Pendulum, has distinct modes that consist of translation of the proof mass along the x and y axis. Axisymmetric gyroscopes, such as ring/disk and wineglass gyroscopes, on the other hand utilize the so-called "wineglass" modes for gyroscope operation. These modes consist of sinusoidal deformations along the radius of the structure and are characterized by a mode number "n" depending on the period of the sinusoid.

For axisymmetric gyroscope applications, such as ring/disk and wineglass gyroscopes, the most commonly used resonance modes are the first two wineglass modes or the so-called $n = 2$ (4-node) and $n = 3$ (6-node) wineglass modes, Figure 7.9. This is due to the fact that lower order wineglass modes have higher angular gain factors and lower resonance frequencies. Each wineglass mode has two degenerate modes that are spaced 45° and 30° apart for $n = 2$ and $n = 3$

7.1 Design of Micro-Glassblown Wineglass Resonators

(a) $n = 2$ wineglass modes   (b) $n = 3$ wineglass modes

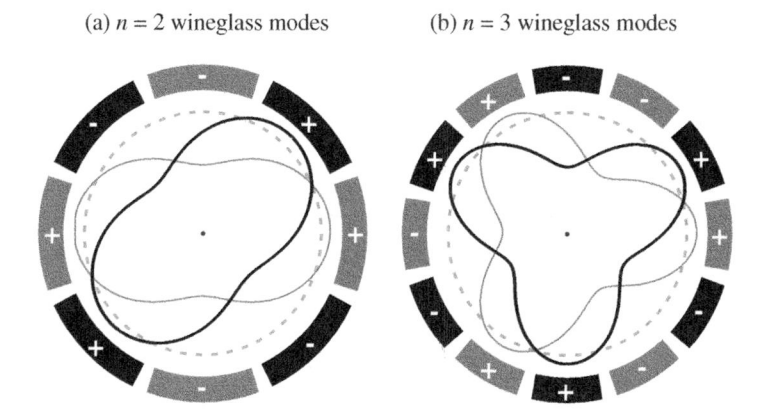

Figure 7.9 Mode shapes and minimal electrode configuration required for $n = 2$ (a) and $n = 3$ (b) wineglass modes. (+) and (−) signs denote in-phase and anti-phase motion, respectively.

modes, respectively. For very low fundamental frequency splits, the degenerate mode pair becomes indistinguishable, becoming a standing wave pattern. Coriolis input into the resonator causes this standing wave pattern to rotate at an angle proportional to angle of rotation (θ), analogous to the Foucualt pendulum.

7.1.2.1 Frequency Symmetry Scaling Laws

Compared to macro-scale HRGs [17], MEMS wineglass resonators have orders of magnitude smaller dimensions, both in shell thickness and diameter. This act of miniaturization requires fabrication processes with very demanding absolute tolerances in order to obtain the required frequency symmetry between the primary wineglass modes. In this section, HRG scaling laws are applied to MEMS sized wineglass resonators to demonstrate the effect of miniaturization.

The geometric imperfections of wineglass resonators can be specified using Fourier series representation of the thickness around the central axis of symmetry, Figure 7.10 [12]:

$$h(\phi) = h_0 + \sum_{i=1}^{\infty} h_i \cos i(\phi - \phi_i), \tag{7.8}$$

where $h(\phi)$ is the thickness of the wineglass resonator along its perimeter and Eq. (7.8) is the Fourier series representation of $h(\phi)$ with respect to azimuth angle ϕ such that: h_0 is the average thickness and h_i is the ith thickness harmonic. This thickness variation will create a corresponding mass variation around the central axis of symmetry of the wineglass resonator according to [12]:

$$M(\phi) = M_0 + \sum_{i=1}^{\infty} M_i \cos i(\phi - \phi_i), \tag{7.9}$$

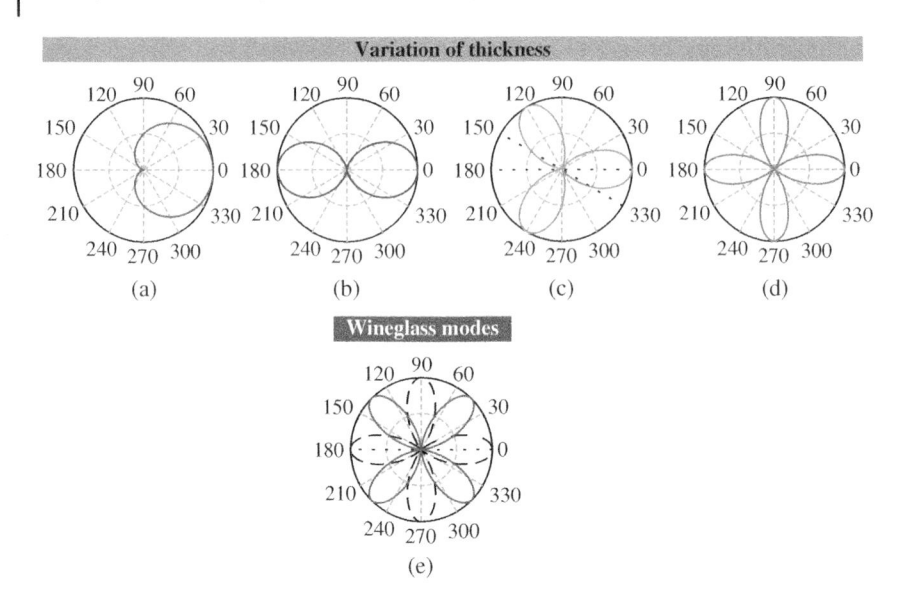

Figure 7.10 Polar plots showing the first four harmonics of thickness imperfections. Only the fourth thickness harmonic affects the frequency symmetry (Δf) of $n = 2$ wineglass modes. (a) First harmonic, (b) second harmonic, (c) third harmonic, (d) fourth harmonic, and (e) $n = 2$ wineglass modes.

where M_0 is the average mass per unit angle and M_i is the ith harmonic of the thickness variation.

For gyroscope applications, the most commonly used resonance modes are the first two wineglass modes or the so called $n = 2$ and $n = 3$ wineglass modes, Figure 7.9. This is due to the fact that lower order wineglass modes have higher angular gain factors and lower resonance frequencies. Each wineglass mode has two degenerate modes that are spaced 45° and 30° apart for $n = 2$ and $n = 3$ modes, respectively. For very low fundamental frequency splits, the degenerate mode pair becomes indistinguishable. Any Coriolis input to the resonator causes the mode shape to rotate at an angle proportional to the angle of rotation. Rate integrating gyroscopes (RIGs) operate by directly measuring this angle.

It has been shown in [137, 138] that a fundamental frequency split in these degenerate modes will be observed only if there is a thickness variation on $i = 4$ or $i = 6$ harmonics, respectively (only if $i = 2n$). This fact makes the wineglass resonators robust to frequency asymmetries. For example, any imperfection in the first, second, or third thickness harmonics will have no effect on the frequency symmetry of the $n = 2$ wineglass mode.

To understand why only the fourth harmonic of the thickness variation has an effect on fundamental frequency splitting (Δf), we look at the vibrational kinetic

Figure 7.11 Sketch of an ideal wineglass (perfectly spherical), showing θ as the angle along the central axis of symmetry and, ϕ as the secondary angle.

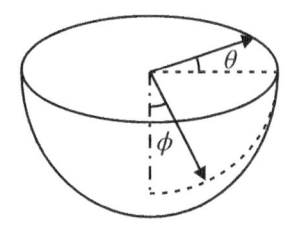

energy of the resonator, which is used in Rayleigh–Ritz solution of the resonance frequencies of wineglass geometries [139, 140]:

$$K_0 = \frac{1}{2}r\rho h \int_0^{\pi/2} \int_0^{2\pi} (\dot{u}^2 + \dot{v}^2 + \dot{w}^2) \sin\phi d\theta d\phi, \tag{7.10}$$

where $\dot{u}, \dot{v}$, and $\dot{w}$ are the velocity terms, which are represented as:

$$u = U(\phi, t) \sin(n\theta), \tag{7.11}$$

$$v = V(\phi, t) \sin(n\theta), \tag{7.12}$$

$$w = W(\phi, t) \sin(n\theta), \tag{7.13}$$

and r is the radius of an ideal wineglass, ρ is the density of the material, θ is the angle along the central axis of symmetry, and ϕ is the secondary angle, Figure 7.11.

However, the above equations for vibrational kinetic energy assume a perfectly symmetric geometry with no thickness variation. If we derive the same equation for an imperfect wineglass resonator with a thickness variation as in Eq. (7.8), then the kinetic energy equation will become:

$$K = \frac{1}{2}r\rho \int_0^{\pi/2} \int_0^{2\pi} (\dot{u}^2 + \dot{v}^2 + \dot{w}^2) h(\phi) \sin\phi d\theta d\phi. \tag{7.14}$$

The difference between kinetic energies of the ideal wineglass resonator in Eq. (7.10) and the one with thickness variations in Eq. (7.14) can be summarized as:

$$K = K_0 + K_{\text{unbalance}}. \tag{7.15}$$

The $K_{\text{unbalance}}$ term, which is the difference in kinetic energy due to thickness variations becomes:

$$K_{\text{unbalance}} = \frac{1}{2}r\rho \int_0^{\pi/2} \int_0^{2\pi} A B \sin\phi d\theta d\phi, \tag{7.16}$$

where A is a collection of velocity terms and:

$$B = h_i \sin(2n\theta) \sin(i\theta). \tag{7.17}$$

Index i is the thickness harmonics under consideration. The B term is the focus of the analysis, as it will make the whole integral (and consequently $K_{\text{unbalance}}$) equal to zero if $2n \neq i$. In other words, only thickness variations with

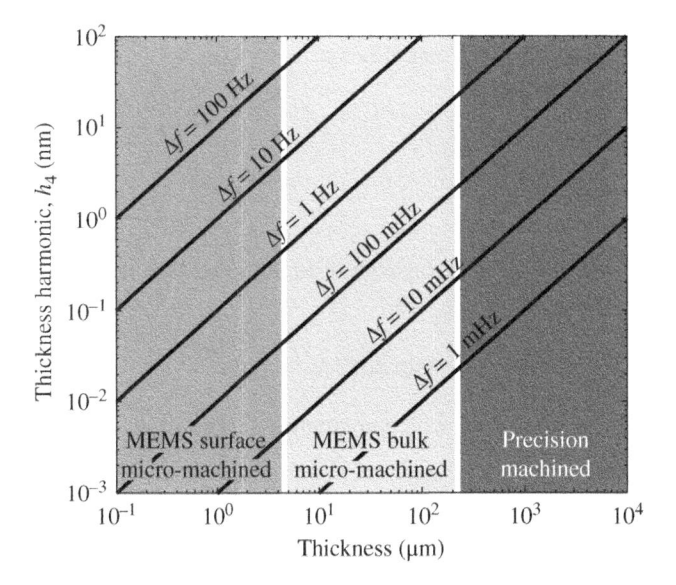

Figure 7.12 Plot showing wineglass thickness versus thickness imperfections in the fourth harmonic and resulting frequency split. Going from precision machined wineglass resonators to micro-machined devices require one to three orders of magnitude improvement in fabrication tolerances due to the reduction in thickness.

harmonics at $2n = i$ can cause changes in the kinetic energy (and frequency) of the wineglass. When the fourth thickness harmonic is not zero, the contribution to the fundamental frequency splitting of $n = 2$ wineglass mode becomes linearly proportional to the fourth harmonic of the shell mass and consequently the shell thickness [137, 138]:

$$\Delta f \cong f \frac{M_4}{M_0} \cong f \frac{h_4}{h_0}. \tag{7.18}$$

Equation (7.18) sets the basis for the scaling laws for frequency imperfections. Because of the thickness term in the denominator, the resonator will become more susceptible to frequency asymmetries as the thickness of the resonator decreases. This effect is shown in Figure 7.12 for a 10 kHz resonator. The thickness axis is divided into three regimes from right to left: macro-scale devices such as the HRGs [17], bulk micro-machined devices which have a thickness range of 10–250 μm, and surface micro-machined wineglass resonators which have a thickness range of 100 nm to 10 μm. As can be seen in Figure 7.12, going from macro-scale devices to MEMS wineglass resonators requires one to three orders of magnitude improvement in absolute tolerances to obtain the same frequency symmetry (Δf).

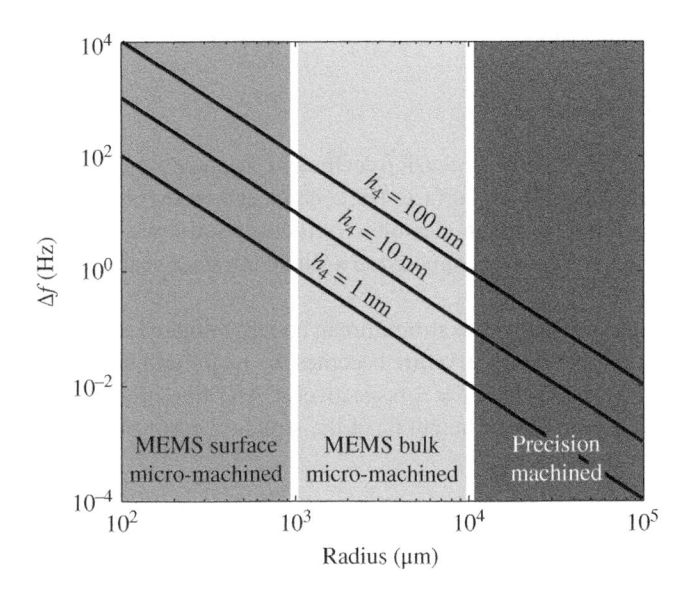

Figure 7.13 The effect of thickness variation of the fourth harmonic on frequency split (Δf) versus wineglass resonator radius.

A similar relationship can be derived for the radius of the wineglass resonator using the Rayleigh–Ritz solution for resonance frequency [140]:

$$f = \sqrt{\frac{n^2(n^2-1)^2 h^2 E}{3(1+\mu)\rho\pi^2 r^4} \frac{\int_0^{\pi/2} \sin\phi \tan\frac{\phi}{2}d\phi}{\int_0^{\pi/2} \{(n+\cos\phi)^2 + 2\sin^2\phi\}\sin\phi d\phi}}. \qquad (7.19)$$

When the above Rayleigh–Ritz solution is inserted into Eq. (7.18), the thickness term in the equation can be eliminated to be replaced with the radius of the wineglass:

$$\Delta f \propto \frac{h_4}{r^2}. \qquad (7.20)$$

The resultant equation shows an even stronger dependence of Δf on radius as the denominator term is squared, Figure 7.13. For every 10× reduction in radius the frequency symmetry deteriorates by 100× 7.13.

7.1.2.2 Stability of Micro-Glassblown Structures

During micro-glassblowing, surface tension forces become active for a brief duration. These forces work toward minimizing the surface energy of the resonator and as a result mitigate the effects of imperfections, such as surface roughness or structural asymmetry. However, if care is not taken, surface tension forces can work toward unbalancing the resonator by creating a pressure instability within the micro-glassblown inverted-wineglass structure.

To analyze this effect we start with the Young–Laplace equation for surface tension:

$$\Delta P = 2\gamma \left(\frac{1}{R_1} + \frac{1}{R_2}\right), \tag{7.21}$$

where ΔP is the pressure, γ is the surface tension coefficient, R_1 and R_2 are the principal radii of curvature of an arbitrary surface. The coefficient 2 on the right-hand side comes from the fact that the micro-glassblown structures have two interface surfaces (inner and outer surfaces), as opposed to a single interface surface such as a droplet of water.

The curvature of an inverted-wineglass structure can be approximated as a hemitoroid where the principal radius of curvature becomes the major and the minor radii of the hemitoroid ($R_1 = R$ and $R_2 = r$, respectively). And the minor radius of the hemitoroid will depend on the height (l) of the structure according to the following geometric expression:

$$r = \frac{l^2 + r_0^2}{2l}, \tag{7.22}$$

where r_0 is the half-width of the trench opening.

Equations (7.21) and (7.22) can be combined to solve for surface tension-induced pressure difference with respect to l. Figure 7.14 shows results of this calculation for inverted-wineglass structures with $R = 1$ mm and $r_0 = 100$–1600 μm, Figure 7.14.

It can be seen that ΔP has a local maxima for all designs, which occur at $r_0 = l$. Interpretation of this result is that the surface tension forces will progressively increase and work toward keeping the structure symmetric if the structure is designed to have $l < r_0$. However, if the shell is glass-blown beyond $l > r_0$, the surface tension forces will progressively decrease. As a result, any perturbation on the geometric shape will be amplified by the further reduction in surface tension-induced pressure ΔP, creating instability within the micro-glassblown structure.

To summarize, in order to achieve high structural symmetry and avoid surface tension-induced instability during micro-glassblowing, inverted-wineglass structures should be designed to have $l < r_0$.

7.2 An Example Fabrication Process for Micro-glassblown Wineglass Resonators

In this section, an example fabrication process for micro-glassblowing of wineglass resonators is reviewed. A typical fused silica micro-glassblowing fabrication process consists of four fundamental steps, namely: (i) etching of the fused silica substrate, (ii) bonding of fused silica (or TSG) device layer to fused silica,

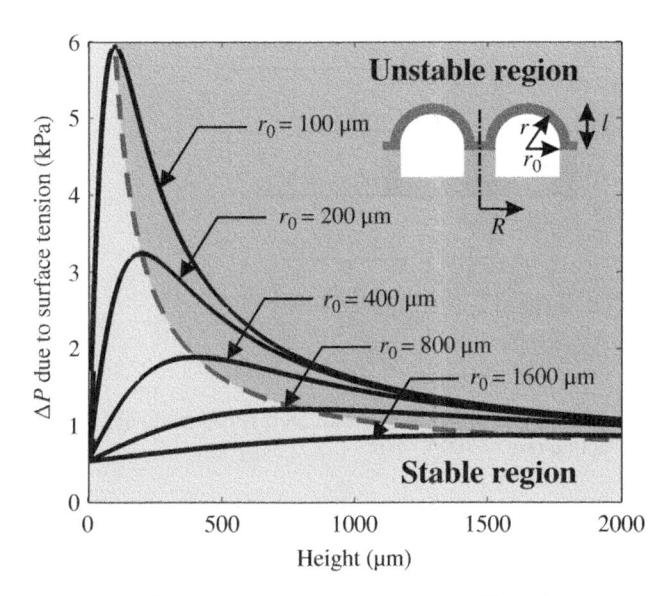

Figure 7.14 Surface tension-induced pressure differential depends on geometric parameters such as cavity radius (r_0) and height (l). $l = r_0$ marks the critically stable region for micro-glassblowing of inverted wineglass structures.

(iii) glassblowing, and subsequently (iv) releasing the wineglass structure by etching around the perimeter, Figure 7.1.

7.2.1 Substrate Preparation

The process starts by LPCVD deposition of a 2 μm polysilicon hard mask onto the fused silica substrate. The cavity openings are defined on the PolySi hard-mask using RIE, followed by wet etching of ~50 μm deep toroidal cavities or 300 μm deep cylindrical cavities into the substrate wafer using concentrated HF (49%).

Once the substrate is etched, PolySi layer is removed via KOH wet etching. Care needs to be taken at this step to remove the PolySi completely, as any residual PolySi can prevent bonding of the device layer. At the same time over-etching the substrate surface using KOH can degrade surface quality of the wafer and reduce bond quality.

7.2.2 Wafer Bonding

The next step of the fabrication process is bonding of the TSG or fused silica device layer onto the pre-etched substrate wafer. Due to the subsequent high temperature glassblowing process the bond needs to survive up to 1700°C, which prevents the use of intermediate materials. For this reason, a plasma activated fusion

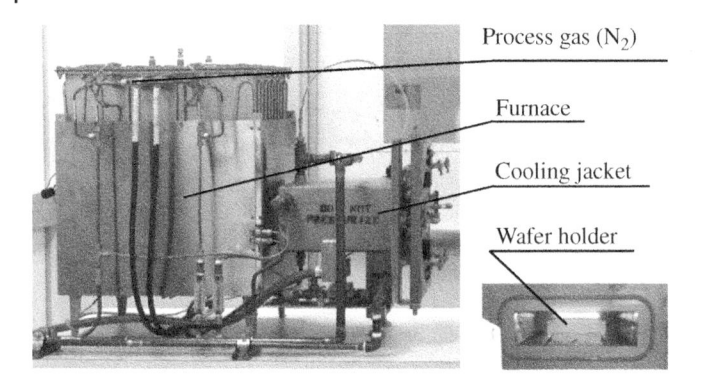

Figure 7.15 Custom-built micro-glassblowing furnace with process capability of 1800°C glassblowing with a rapid cooling rate of 500°C/min.

bonding process is used, Figure 7.1b. The bond is performed by plasma activating the TSG/fused silica wafers and then bringing them into optical contact [141].

Plasma-assisted fusion bonding works by creating hydrogen bonds between the device and the substrate wafers; it requires highly polished, flat, clean surfaces (<10 nm Sa roughness). The process for bonding fused silica or TSG wafer pairs can be divided into four main steps:

1) Cleaning of the wafer pair using solvent and RCA clean,
2) Plasma activation using oxygen plasma,
3) DI water rinse followed by N_2 dry,
4) Optical contact bonding of the activated surfaces,
5) Curing the wafer stack at 400°C for >6 h.

Once cured, the bond creates a seamless hermetic seal around the etched cavities without using any intermediate material.

7.2.3 Micro-Glassblowing

The TSG/fused silica wafer stack is then glassblown at 1600–1700°C in a custom-designed high temperature furnace with a rapid cooling rate of 500°C/min, Figure 7.15. The furnace consists of two main chambers that are connected to each other through a third vestibule chamber. The first chamber is used for heating and can go up to 1800°C; the second chamber is enveloped by a water-cooled jacket that maintains <200°C temperature. The samples are transferred between the heating and cooling chambers by using a sliding alumina wafer holder. A typical glassblowing run involves keeping the wafer stack at glassblowing temperature for one minute and then extracting the wafer stack into the water cooled jacket for solidification.

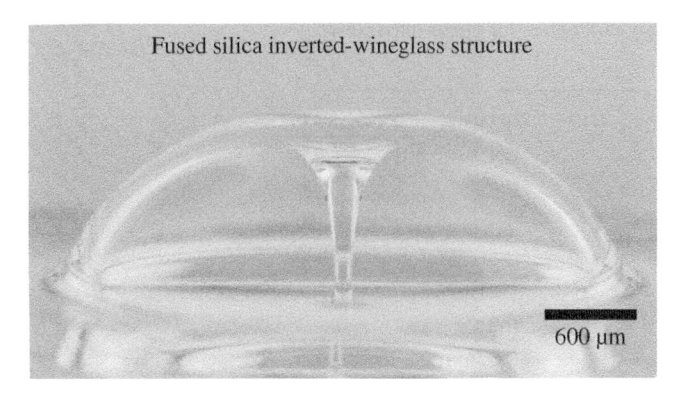

Figure 7.16 Optical photograph of glassblown fused silica inverted-wineglass structure. Outer diameter is 4200 μm. Glassblown at 1650°C.

Figure 7.17 Optical photograph of fused silica spherical shell structures, glassblown at 1700°C. Outer diameter of shells is 800 μm [13]. © IEEE.

During glassblowing, as previously described, two phenomena occur simultaneously: device layer becomes viscous due to the elevated temperature and the air inside the etched cavity expands, creating the 3-D glassblown structure. Because the device layer is bonded both around the circular cavity and the cylindrical post in the middle, the glassblown structure creates a self-aligned stem, Figure 7.16. Whereas if no central post is defined, the glassblown structure forms a spherical shell as in Figure 7.17. Typical design parameters for a mushroom and hemispherical shell structures can be found in Tables 7.3 and Table 7.4, respectively.

7.2.4 Wineglass Release

The final step of the fabrication process is to release the wineglass around its perimeter, which can be accomplished by laser ablation, lapping, or dry/wet

Table 7.3 Sample design parameters for micro-glassblown mushroom structure.

Device diameter	(μm)	3800
Central post diameter	(μm)	400
Device layer thickness	(μm)	300
Etch depth	(μm)	250
Final height	(μm)	900
Glassblowing temperature	(°C)	1700
Glassblowing time	(min)	3
Cooling rate	(°C/min)	500

Table 7.4 Sample design parameters for micro-glassblown hemisphere.

Device diameter	(μm)	900
Device layer thickness	(μm)	200
Etch depth	(μm)	350
Final height	(μm)	900
Final equator diameter	(μm)	1200
Glassblowing temperature	(°C)	1700
Glassblowing time	(min)	3
Cooling rate	(°C/min)	500

etching of the device layer. The wineglass structure shown in Figure 7.18 was released with laser ablation, using a 2-axis laser micro-machining system, Resonetics RapidX 250. 3-Axis laser ablation capability was added by implementing a custom built rotary stage assembly from National Aperture, Inc. In order to laser ablate around the perimeter of the wineglass, the structure was mounted onto the rotary stage and its axis of symmetry was aligned with the rotation axis using an x–y stage. Laser ablation was performed by focusing the laser beam onto the perimeter of the wineglass at a perpendicular angle and rotating the wineglass structure at constant angular velocity.

7.3 Characterization of Micro-Glassblown Shells

In this section, the surface roughness and the material composition of micro-glassblown TSG device layer before and after glassblowing is reviewed.

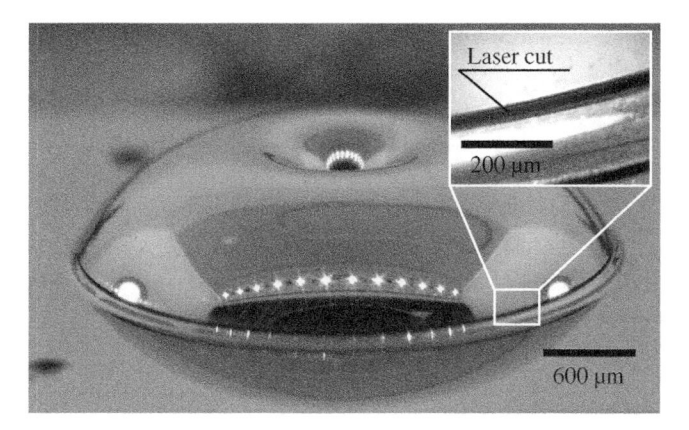

Figure 7.18 Optical photograph of inverted-wineglass, released along the perimeter using laser ablation and coated with iridium [13]. © IEEE.

7.3.1 Surface Roughness

In order to minimize the surface losses in resonant and optical applications, highly smooth surfaces are required. Surface roughness measurements of TSG glassblown samples were performed using an atomic force microscope (AFM) from Pacific Nanotechnology (Nano-R). With a sensor noise level of <0.13 nm in the z-direction, Nano-R can resolve sub-nanometer features. Samples were cleaned using standard solvent clean (acetone, IPA, methanol) before each scan. No additional treatment was performed on the samples. The AFM was run in tapping mode, using a <0.13 nm radius probe tip (Agilent U3120A).

Surface roughness of the samples before and after glassblowing at 1600°C were analyzed, with the hypothesis that glassblowing can improve the surface roughness. Highly polished TSG wafers were used for the device layer, which was verified by AFM scans, showing a surface roughness of 0.40 nm Sa, Figure 7.19b. Characterization of the glassblown samples showed a two-fold improvement in surface roughness, down to 0.23 nm Sa, Figure 7.19a. We also observed that the angstrom level scratches in Figure 7.19b, associated with the lapping operation, disappeared after glassblowing, Figure 7.19a, confirming the hypothesis.

Two-fold improvement in surface roughness is attributed to viscous flow of the glass layer and the associated surface tension forces. As the glassblowing is performed above the glass softening temperature, TSG device layer becomes viscous and the surface tension forces become active, working toward minimizing the surface area of the glass structure. This creates an effect analogous to "stretching out" the wrinkles on the surface, lowering the surface roughness.

It was later shown in [142] that a substantially lower thermal re-flow temperature of 1300°C is sufficient to achieve a significant reduction in surface roughness.

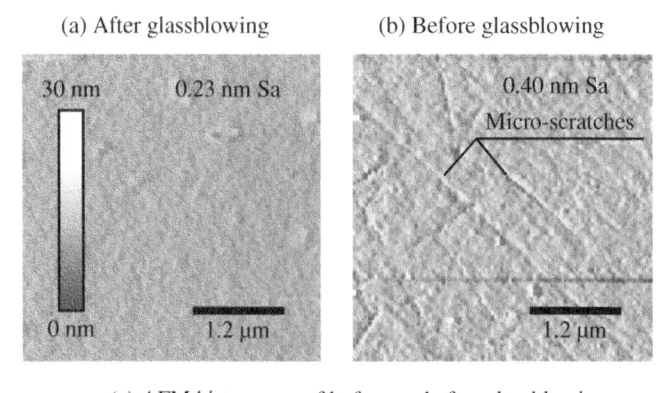

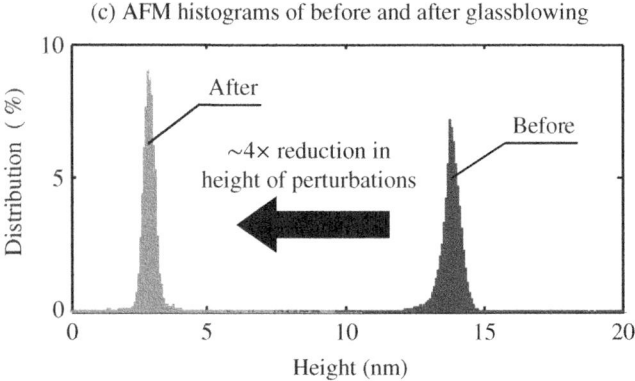

Figure 7.19 AFM surface profiles of TSG, (a) before and (b) after glassblowing, showing ~4x reduction in surface roughness (c). Glassblowing creates extremely smooth (0.2 ppm relative roughness) TSG structures.

7.3.2 Material Composition

For resonant and optical applications, it is critical that TSG retains its original material composition and properties after glassblowing, which are structural integrity, material uniformity, and optical transparency. We found that glassblowing temperature and the rate of cooling are the most important parameters that affect the quality of the TSG layer after glassblowing.

The structure in Figure 7.20 was glassblown using a conventional high temperature furnace at 1600 °C, which does not allow removal of the samples at elevated temperatures. For this reason, the structure was left to cool-down to room temperature over an 8 h period. Recrystallization as well as micro-cracks were observed on the surface. In order to establish the nature of the recrystallization, electron dispersive spectroscopy (EDS) was employed. Philips XL-30 FEG SEM with a Thermo Scientific UltraDry silicon drift X-ray detector was used for EDS characterization.

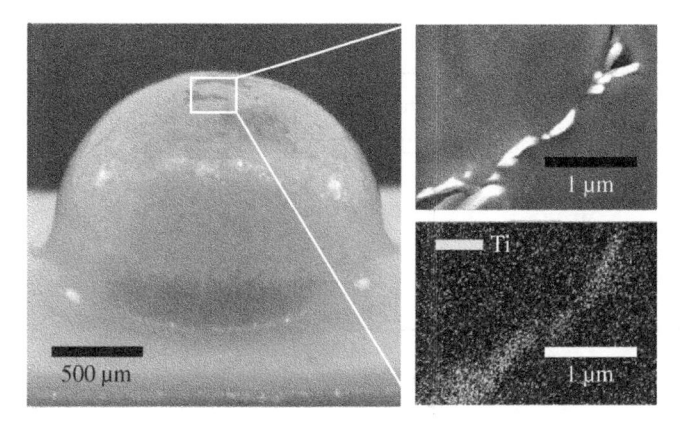

Figure 7.20 Slow cooling of TSG (>8 h) causes recrystallization [13]. © IEEE.

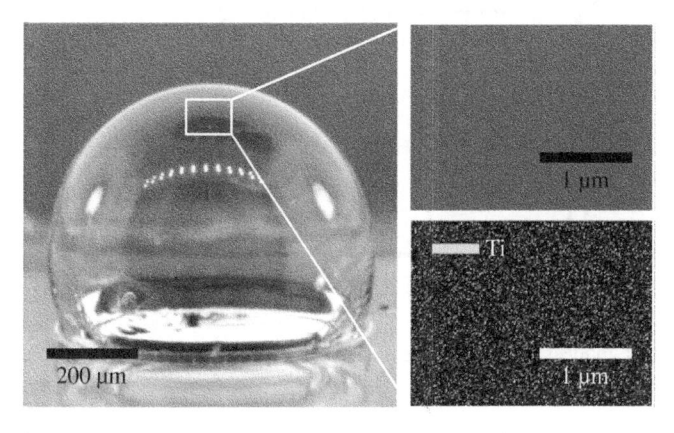

Figure 7.21 Glassblowing with rapid cooling of TSG (< 1 min) prevents recrystallization [13]. © IEEE.

An acceleration voltage of 10 kV was used at 10 mm working distance, and samples were coated with 5 nm of sputtered iridium to prevent charging.

EDS analysis of the crystals in Figure 7.20 revealed higher concentrations of titanium, implying that TiO_2 is exsolving from the SiO_2/TiO_2 matrix. In contrast, the structure in Figure 7.21 was glassblown using rapid cooling by bringing the temperature of the sample from 1600 – 1700°C to ~200 °C within a minute. No micro-cracks or recrystallization were observed, as can be validated by the optical transparency of surfaces. The EDS spectral plots showed homogeneous SiO_2 and TiO_2 distribution in Figure 7.21 as opposed to heterogeneous distribution in Figure 7.20. The absence of recrystallization makes rapid cooling an essential step in micro-glassblowing of fused silica and TSG.

(a) EDS spectral analysis of ULE TSG

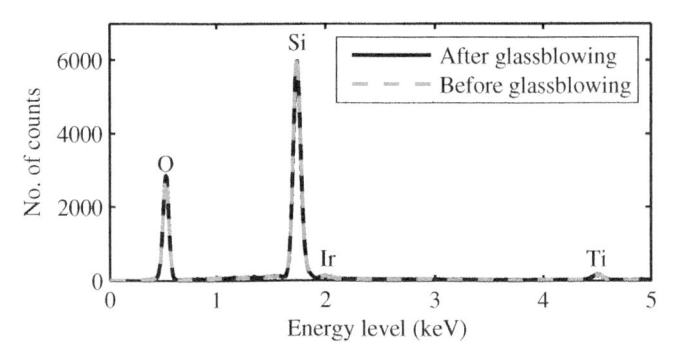

(b) EDS spectral analysis of fused silica

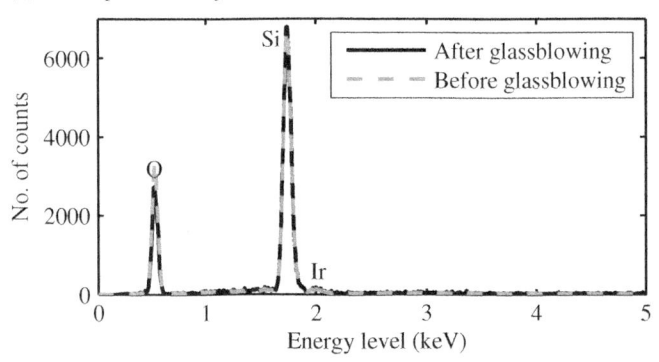

Figure 7.22 EDS spectral analysis of TSG and fused silica reveals that composition of the material does not change after glassblowing.

EDS was used to obtain the spectral signatures of TSG before and after glassblowing. No change in the composition of TSG was observed after glassblowing, Figure 7.22. EDS spectrum also revealed 7–8 wt % of TiO_2 in TSG, which is in agreement with the nominal TiO_2 concentration of Corning ULE TSG.

8

Transduction of Micro-Glassblown Wineglass Resonators

In this chapter, three different electrode architectures for electrostatic transduction of micro-glassblown wineglass resonators are reviewed. These are: (i) assembled electrodes for preliminary characterization of micro-glassblown wineglass resonators, (ii) in-plane electrodes, and (iii) out-of-plane electrodes.

8.1 Assembled Electrodes

In Chapter 6, a review of various micro-glassblown wineglass resonators was presented; in Chapter 7, micro-glassblowing of fused silica for fabrication of micro-wineglass structures was reported. Despite these recent advances in micro-wineglass resonator fabrication, the technology development is in its infancy and characterization of the early stage prototypes presents a challenge. This section introduces temporary electrode structures for characterization of micro-wineglass devices of any shape or size, eliminating the need for in situ electrode fabrication during the development cycle of the resonator architecture.

8.1.1 Design

In one possible implementation of electrodes, electrostatic excitation and detection are provided by assembled electrode structures, which are fabricated separately from the micro-glassblown resonator on an SOI wafer, Figure 8.1. In such implementation, the SOI stack consists of a 500 µm silicon substrate layer, a 5 µm buried oxide layer, and a 100 µm silicon device layer. Each electrode assembly consists of eight independent electrodes that are spaced 45°C apart around a central thru-hole. Large spring structures on each electrode allow an adjustment distance up to 400 µm for each individual electrode (total of 800 µm along the resonator

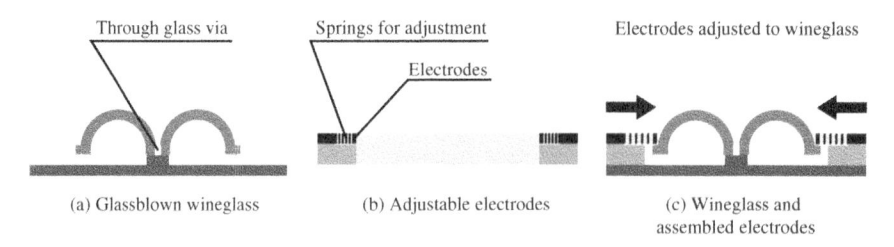

Figure 8.1 Electrodes are fabricated separately on an SOI stack, bonded to the resonator wafer and extended.

(a) Glassblown wineglass
(b) Adjustable electrodes
(c) Wineglass and assembled electrodes

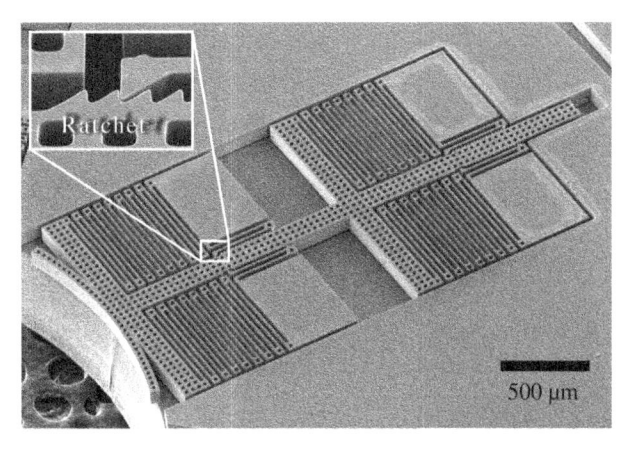

Figure 8.2 SEM image of an adjustable electrode with 400 µm maximum displacement and 10 µm positioning resolution [143]. © IEEE.

diameter), which permits a single electrode design to be used for different wineglass architectures and diameters, Figure 8.2. Electrode assemblies with seven different central-hole diameters were fabricated on the same SOI wafer, which covers all wineglass diameters from 1 to 4.5 mm.

8.1.2 Fabrication

The fabrication of SOI electrodes starts by lithographically patterning the back-side thru-hole and etching using DRIE to a depth of 500 µm. This is followed by patterning and etching the device layer to create the electrode structures. For the electrode structures, the DRIE is performed to a depth of 100 µm, using a 1.6 µm oxide hard mask for better feature resolution, then O_2 ashing, followed by RCA-1 cleaning is used to remove the excess photoresist and etch residues. The final step of the fabrication process is HF wet etching of buried oxide layer to release the electrode structures.

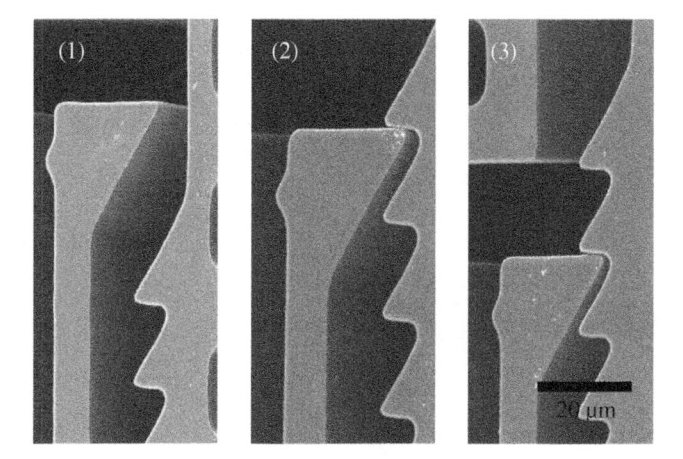

Figure 8.3 Ratchet mechanism acting on the electrode structure, the electrode is extended gradually in 1 through 3 [143]. © IEEE.

8.1.2.1 Experimental Characterization

Once the fabrication is complete, the electrode structures are singulated and bonded onto the micro-wineglass die, Figure 8.1. Then, each electrode is pushed to close proximity of the wineglass by using a micro-manipulator. When the correct location is achieved, four ratchet mechanisms (two front and two back) keep the electrodes in place, Figure 8.3. The ratchet mechanisms act on two rack gears placed on the electrode. The pitch distance on the rack gears are 20 µm. By offsetting the front and rear ratchet mechanisms by 10 µm relative to the teeth pitch of the rack gear, a positioning resolution of 10 µm was obtained (minimum capacitive gap).

For experimental characterization, assembled electrode structures were bonded onto micro-glassblown wineglass structures, Figure 8.4. A wineglass with 4.2 mm diameter, 50 µm thickness, and 300 µm central stem was tested using an assembled electrode structure with 4.5 mm thru-hole. The entire assembly was placed into a ceramic DIP package with a gold back-plate and wirebonded at the anchors of each electrode, Figure 8.5. The bias voltage to the resonator was applied through the gold back-plate, which connects to the resonator metal layer through the via at the center of the stem.

Frequency sweeps were obtained using an Agilent 4395A network analyzer. Two opposite electrodes were used with the goal of forced excitation of the $n = 2$ wineglass mode. A DC voltage of 10 V and an AC voltage of 5 V were used. A Q-factor of 40.000 was observed at 14.8 kHz giving a time decay constant of ~0.9 s, Figure 8.6.

Adjustable nature of the assembled electrode structures allows characterization of 3-D resonator structures with varying size and diameter.

114 | *8 Transduction of Micro-Glassblown Wineglass Resonators*

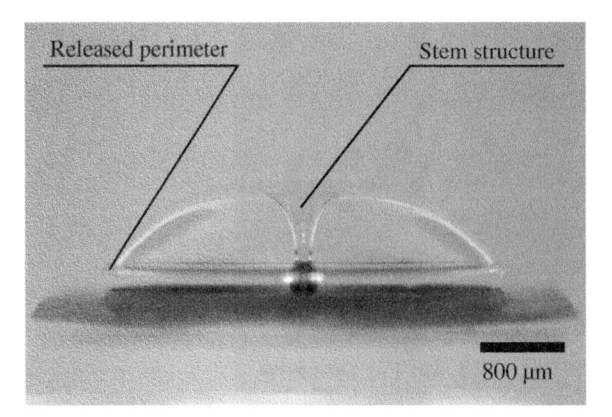

Figure 8.4 Released wineglass structure with 4.2 mm diameter, 50 μm thickness, and 300 μm central stem.

Figure 8.5 Electrode structures assembled onto a micro-glassblown wineglass resonator with <20 μm gaps.

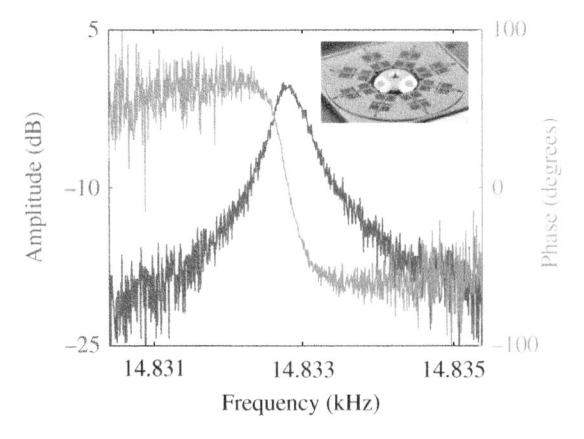

Figure 8.6 Electrostatic frequency sweep using adjustable electrode assembly, showing $Q = 40$k at 14.8 kHz.

Figure 8.7 A glassblown spherical resonator with assembled electrodes. Diameter is 1.2 mm and thickness is 5 μm [143]. © IEEE.

Figure 8.7 shows assembled electrodes structures around a 1.2 mm micro-glassblown spherical resonator with resonant frequencies in the MHz range.

8.2 In-plane Electrodes

In this section, micro-glassblowing process is expanded to include co-fabrication of in-plane electrode structures.

8.3 Fabrication

In order to fabricate the micro-wineglass resonators with in-plane electrodes, first cylindrical cavities with a central post were etched to 250 μm depth on a 100 mm silicon substrate wafer using DRIE, Figure 8.8a. Then, a thin glass layer (100 μm) was anodically bonded to the silicon substrate. Anodic bonding was performed using a DC voltage of 600 V and a load of 100 N at 400°C. The glass layer was bonded to the substrate along the perimeter of the cylindrical cavity and at the central post, hermetically sealing atmospheric pressure air within the cavities. This was followed by deep glass dry etching to define the outer perimeter of the wineglass resonator and the central via hole, Figure 8.8b. Capacitive gaps and individual electrodes as well as the central via hole were defined at this step. The glass etching was performed using a magnetic neutral loop discharge plasma oxide etcher (ULVAC NLD 570 Oxide Etcher) [144]. A ~5 μm thick low-stress electroplated Cr/Ni hard-mask was used to etch the 100 μm deep trenches. This was followed by micro-glassblowing of the wafer stack at 875°C inside an RTA system, where the glass layer becomes viscous and the air inside the cavity expands, creating the 3-D shell structure, Figure 8.8c. Once the 3-D micro-glassblown structure forms, the wafer was rapidly cooled to room temperature for solidification.

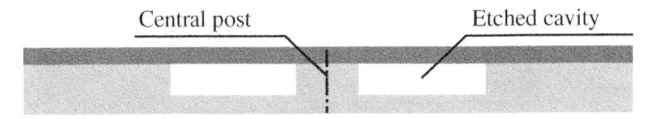

(a) Silicon substrate is etched and glass device layer is bonded

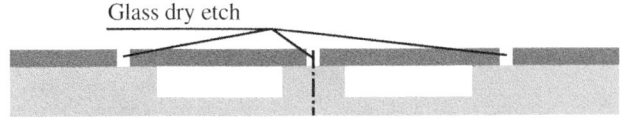

(b) Glass layer is etched, defining the perimeter and electrodes

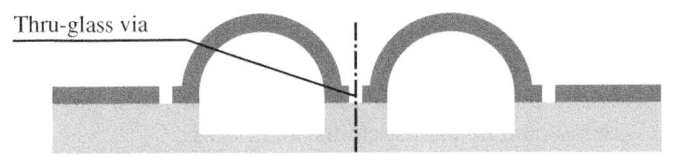

(c) Wafer stack is glassblown creating the 3-D shell structure

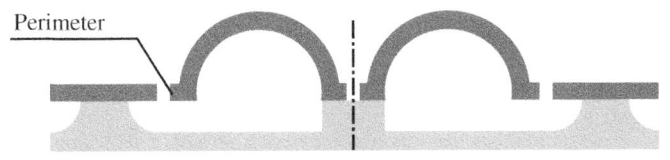

(d) Silicon is etched using XeF_2 to release the wineglass structure

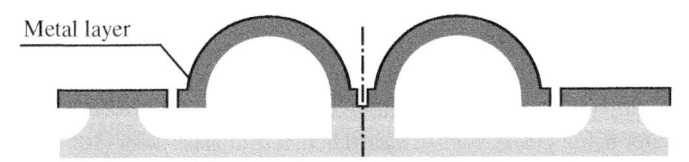

(e) A thin metal layer is blanket coated using sputtering

Figure 8.8 Process flow for fabrication of micro-glassblown wineglass resonators with integrated electrodes. (a) Device layer is bonded to pre-etched substrate, (b) device layer is etched, defining the perimeter of the wineglass and the electrodes, (c) wafer stack is glassblown, (d) silicon underneath the perimeter is etched using XeF_2, and (e) a thin metal layer is blanket coated.

During the micro-glassblowing step, the perimeter of the wineglass structure and the planar electrodes do not deform as there is no etched cavity under these structures, enabling lithographic definition of the capacitive gaps. The next step was XeF_2 etching of the substrate underneath the glass layer in order to release the wineglass resonator along its perimeter, Figure 8.8d. XeF_2 was chosen because of the extremely high glass to silicon selectivity (as high as 1 : 1000 selectivity).

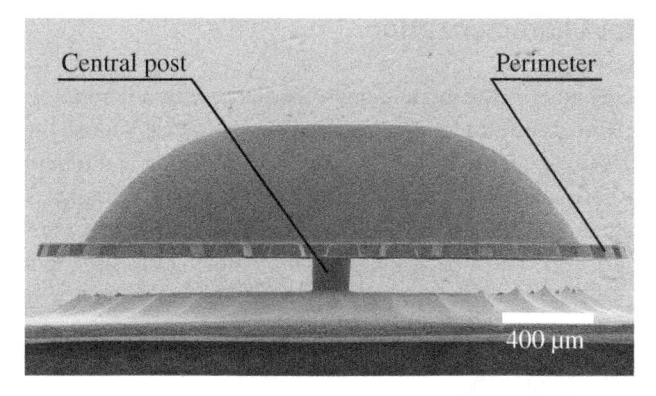

Figure 8.9 SEM image of a stand-alone micro-wineglass structure after release. Diameter 4.4 mm, thickness 50 µm [145]. © IEEE.

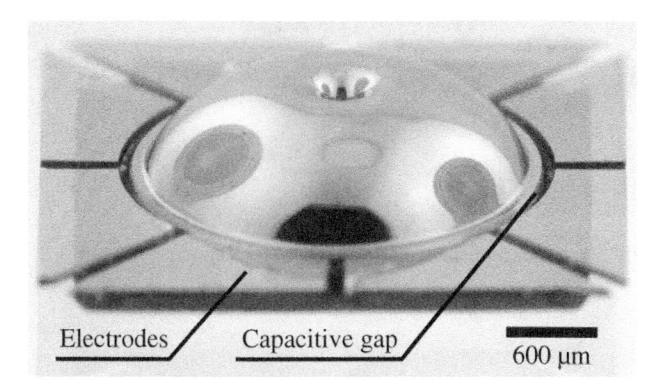

Figure 8.10 Metallized micro-wineglass structure with integrated electrodes. Diameter 4.4 mm, thickness 50 µm.

Once the etch was complete, a free standing micro-wineglass structure with a self-aligned stem structure was obtained, Figure 8.9.

Final step of the fabrication process is blanket metallization by sputtering, Figure 8.10. A 30 nm sputtered Iridium layer was chosen for the metal layer, because of high conductivity, corrosion resistance, and the ability to apply to the surfaces without utilizing an adhesion layer (such as Cr or Ti). The metal layer coats the top surface of the resonator shell, the side walls of the capacitive gaps, as well as inside of the central via hole. However, directionality of the sputtering process prevents the metal layer from coating the undercut created by the XeF_2 etch, electrically isolating the electrodes and the resonator, Figure 8.8e. Electrical feed-through to the resonator was obtained through the central via structure, which connects the resonator to the substrate, Figure 8.11.

8.4 Experimental Characterization

In order to experimentally identify the mode shapes associated with different resonant frequencies, the wineglass resonator was excited electrostatically using the integrated electrode structures, Figure 8.11. The amplitude of motion at different points along the outer perimeter was mapped using Laser Doppler Vibrometry (LDV), creating a representation of the mode shapes associated with different resonant frequencies, Figure 8.12. This was accomplished by moving the laser spot

Figure 8.11 Packaged and wirebonded micro-wineglass resonator. Diameter 4.4 mm, thickness 50 μm

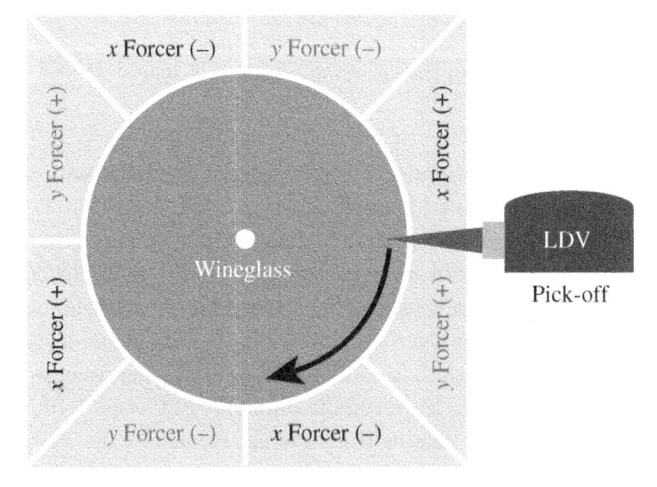

Figure 8.12 Laser Doppler Vibrometer was used to scan along the perimeter of the wineglass to map the mode shapes associated with x forcer and y forcer electrodes.

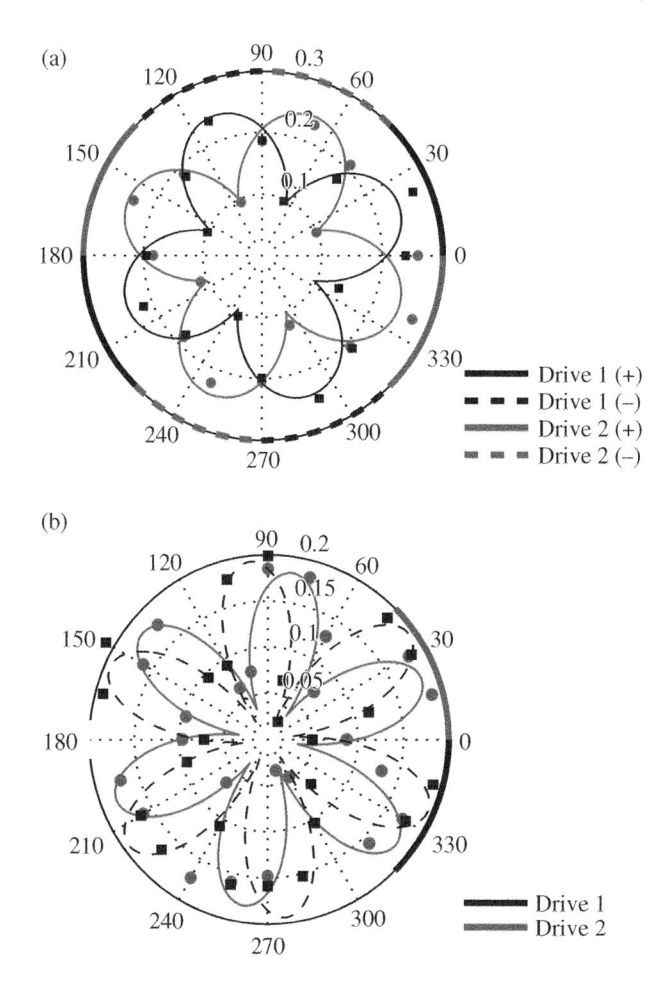

Figure 8.13 Measured velocity amplitude distribution (mm/s) identifying (a) $n = 2$ and (b) $n = 3$ wineglass modes.

along the perimeter while driving the resonator with two different sets of electrode configurations for each degenerate wineglass mode, Figure 8.13.

For $n = 2$ wineglass mode, four electrodes were used for each degenerate mode with $45°$ angle between the two electrode sets. Two of the electrodes were driven in anti-phase. This electrode configuration excites the $n = 2$ wineglass mode selectively, while suppressing all other modes. For $n = 3$ mode, a single electrode were used for each degenerate wineglass mode. Excitation using a single electrode was necessary, as a balanced excitation using two or four electrodes inherently suppresses the $n = 3$ mode. A DC bias voltage of 100 V and an AC drive voltage of 5 V

Table 8.1 Table summarizing frequency splits and center frequency of five different micro-wineglass structures.

Device no.	Center freq. (Hz)	Δf (Hz)	σ (Hz)	$\Delta f/f$ (ppm)
1	27388.65	0.16	0.04	5.67
2	28889.18	4.69	0.05	162.18
3	29227.60	1.76	0.05	60.30
4	29090.38	21.08	0.06	724.65
5	29442.98	9.61	0.07	326.40

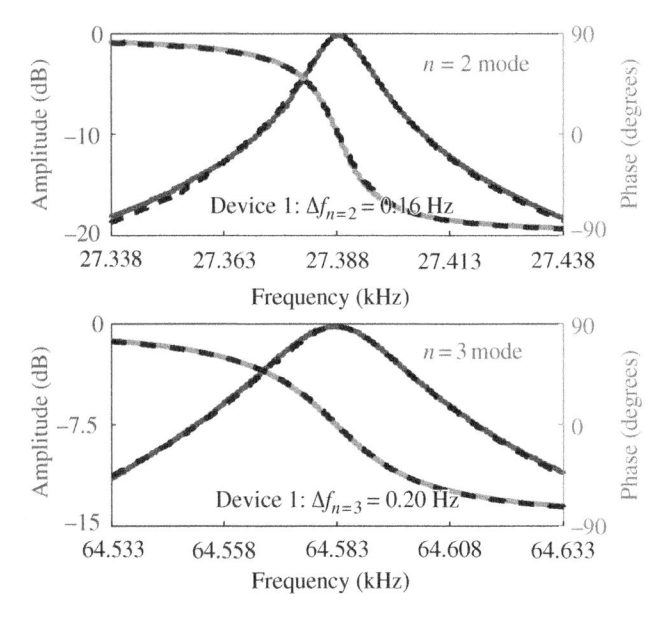

Figure 8.14 Experimental frequency sweeps of $n = 2$ and $n = 3$ wineglass modes, showing $\Delta f = 0.16$ Hz and $\Delta f = 0.20$ Hz, respectively.

was used in all experiments. Large drive voltages used in this experiment were due to large capacitive gaps of the current prototypes (>30 µm).

For device 1, Table 8.1 center frequencies of degenerate wineglass modes were identified at 27389 and 64583 Hz for $n = 2$ and $n = 3$ wineglass modes, respectively, Figure 8.14. Frequency splits between the two degenerate modes were measured by fitting a second-order system response onto the frequency sweep data of each degenerate wineglass mode. For the device 1 highlighted in these measurements, the frequency splits (Δf) of 0.15 and 0.2 Hz were observed for $n = 2$ and $n = 3$ wineglass modes with 95% confidence levels at 0.23 Hz for

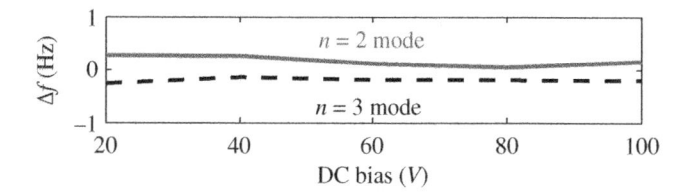

Figure 8.15 Frequency split versus DC bias, showing that the frequency split is within 1 Hz independent of DC bias (DC bias was varied between 20 and 100 V with 20 V increments).

$n = 2$ and 0.3 Hz for $n = 3$, Figure 8.14. In order to estimate the contribution of electrostatic spring softening effect, DC bias voltage was varied between 20 and 100 V, frequency split stayed below 1 Hz for both modes, attributing the low frequency split to high structural symmetry and not to capacitive tuning, Figure 8.15.

In order to verify the repeatability of the results, four other wineglass resonators were characterized using the same method described above. Three of the five wineglass resonators had frequency split less than 5 Hz, one less than 10 Hz for the $n = 2$ wineglass mode, with one outlier at $\Delta f = \sim 21$ Hz, Figure 8.16.

Identification of the mode shapes using Laser Doppler Vibrometry revealed frequency splits as low as $\lesssim 1$ Hz at ~ 27 kHz center frequency on device #1, giving a relative frequency split of $\Delta f_{n=2}/f_{n=2} < 10$ ppm (or 0.001%). Three of the five wineglass resonators had frequency splits less than 5 Hz, one less than 10 Hz for the $n = 2$ wineglass mode, with one outlier at $\Delta f = \sim 21$ Hz, Table 8.1.

The focus of this study was on frequency symmetry of micro-glassblown resonators, and for this reason borosilicate glass was used as the resonator material. As expected, low Q-factors (several thousands) were observed due to the high internal dissipation of borosilicate glass. Also, large capacitive gaps were used for electrostatic transduction, due to challenges associated with deep glass dry etching, which required use of high DC bias voltages for excitation. Future research directions could include high-Q materials such as ULE TSG/fused silica for the resonator material as well as smaller capacitive gaps in order to achieve high performance rate-integrating gyroscope (RIG) operation. ULE TSG/fused silica glassblowing process previously demonstrated by the authors in [13, 146] and a high temperature substrate such as tungsten may help achieve the Q-factors required for RIG operation. Improved dry etching performance of 7 : 1 aspect ratio also demonstrated by the authors [144], coupled with a thinner device layer, is expected to provide smaller capacitive gaps and improved electrostatic transduction.

These results demonstrate the feasibility of surface tension driven microglassblowing process as a means to fabricate extremely symmetric and smooth 3-D wineglass resonators. High structural symmetry $\Delta f < 1$ Hz and atomically

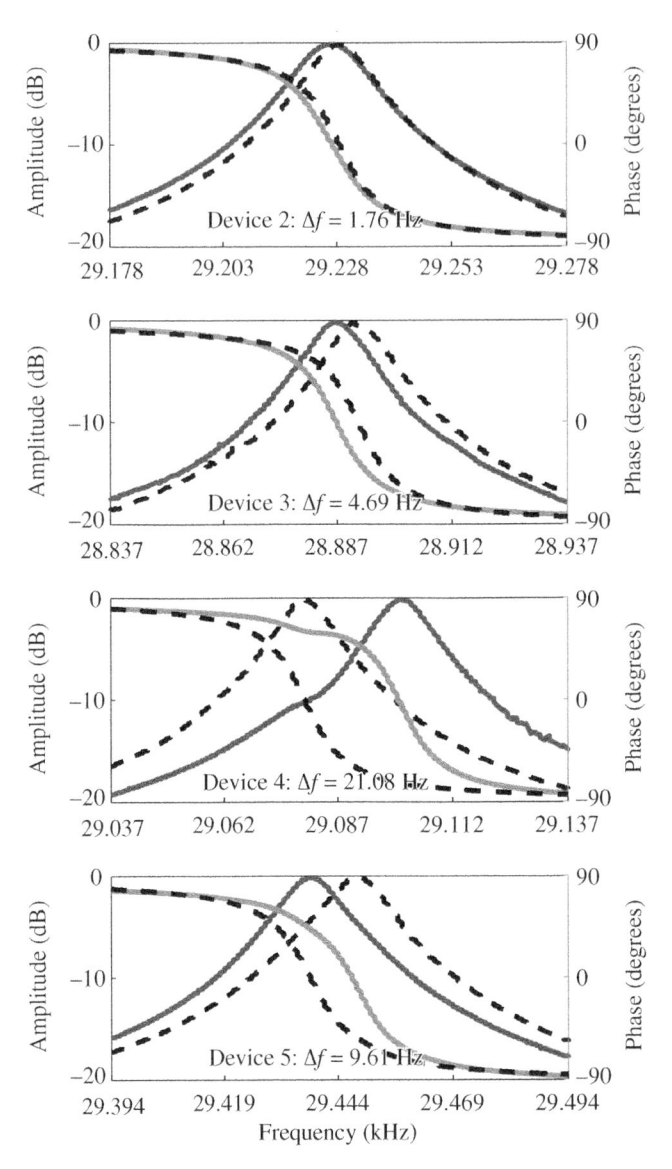

Figure 8.16 Frequency sweeps of $n = 2$ mode of four additional wineglass resonators, showing Δf values in the range of 1.76 – 21.08 Hz.

smooth surfaces (0.23 nm Sa) of the resonators may enable new classes of high performance 3-D MEMS devices, such as rate-integrating MEMS gyroscopes and mode-matched angular rate gyroscopes.

8.5 Out-of-plane Electrodes

An alternative transduction scheme relies on detection of spatial deformation of the 3-D wineglass structure using an out-of-plane electrode structures. In addition to the potential advantages in simplicity and ease of implementation, out-of-plane electrode architecture enables the use of sacrificial layers to define the capacitive gaps.

8.6 Design

Wineglass Coriolis Vibratory Gyroscopes (CVGs) typically utilize eight or more electrodes to drive and sense the primary wineglass modes. One of the main challenges of fabricating micro-wineglass resonators is the definition of electrode structures in a manner compatible with batch fabrication, Figure 8.17. 3-D side walls of the wineglass geometry makes it challenging to fabricate radial electrodes with small capacitive gaps and to keep the gap uniform across the height of the structure. Even though postfabrication assembly techniques have been successfully demonstrated [103, 147], these approaches create a bottle-neck in batch-fabrication of the devices at the wafer level.

In this chapter, we explore an alternative transduction paradigm based on out-of-plane electrode architecture. The architecture consists of a microglassblown fused silica wineglass resonator and planar Cr/Au electrodes defined on a fused silica substrate, Figure 8.18. Out-of-plane capacitive gaps are formed between the Cr/Au metal traces and the perimeter of the wineglass resonator. Electrostatic transduction is made possible by the 3-D mode shape of the wineglass resonator. In-plane deformation of wineglass modes is accompanied by an out-of-plane deformation, Figure 8.19. This permits the use of out-of-plane transduction to drive and sense the in-plane oscillations, which are sensitive to Coriolis forces along the z-axis of the structure [92].

In our implementation, a total of eight electrodes are used, which is the minimal configuration to drive and sense the $n = 2$ wineglass modes. Four electrodes are designated as forcer (FX and FY) and Four are designated as pick-off (PX and PY). Both the forcer and pick-off channels have differential pairs (i.e. FX+ and FX−). The resonator is biased using any of the eight traces that extend from the central

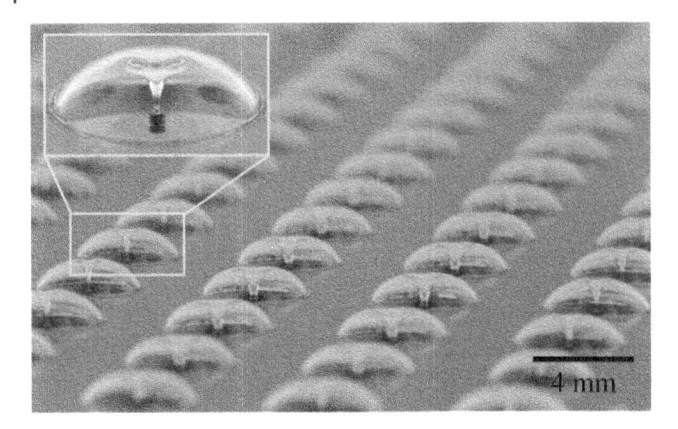

Figure 8.17 Micro-glassblowing process can create arrays of inverted-wineglass structures on the wafer surface. Outer diameter of shells is ~4 mm, over 100 Borosilicate Glass (BSG) shells were fabricated on a 100 mm wafer.

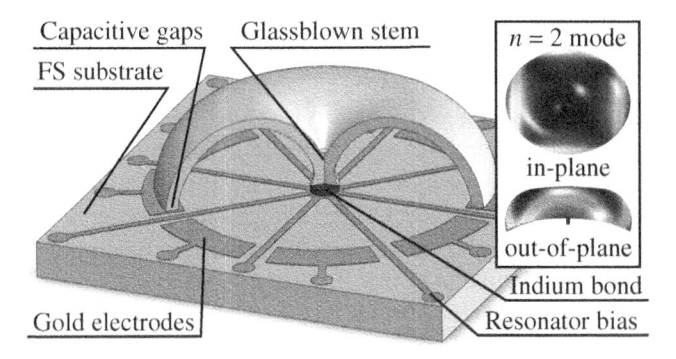

Figure 8.18 Out-of-plane electrode architecture consists of a micro-glassblown fused silica (FS) wineglass resonator and planar Cr/Au electrodes defined on fused silica, enabling batch-fabrication.

anchor point. These traces also help suppress parasitic coupling between adjacent electrodes by sinking stray currents, Figure 8.20.

As the thickness of the shell limits the maximum surface area for the out-of-plane electrodes, typically a smaller surface area is utilized for capacitive gaps compared to radial electrodes. However, this is offset by the fact that the electrode structure is of planar nature, which makes it easier to obtain smaller capacitive gaps and helps to compensate for the loss of surface area. In addition, sacrificial layers and wafer-to-wafer bonding techniques can be used to define the capacitive gaps, which makes the process very robust to alignment errors, as the gap uniformity is defined by the thickness of the sacrificial layer and not

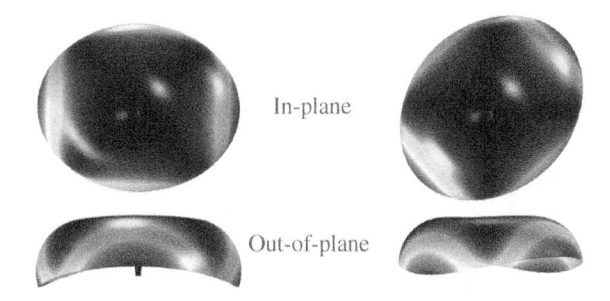

(a) First $n = 2$ wineglass mode (b) Second $n = 2$ wineglass mode

Figure 8.19 Out-of-plane transduction scheme utilizes out-of-plane component of wineglass modes to drive and sense in-plane motion.

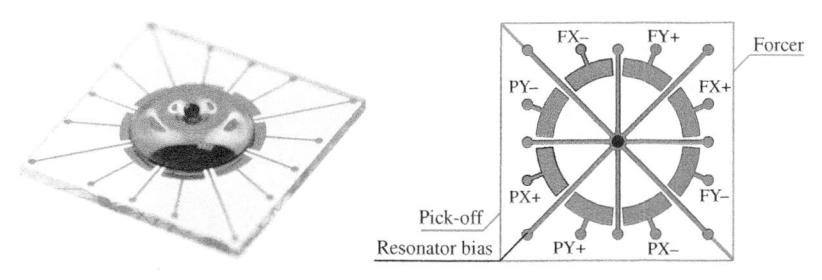

(a) Fused silica wineglass with out of plane electrodes. Shell diameter is 3.8 mm

(b) Schematic of the out-of-plane electrodes, showing resonator bias, pick-off and forcer electrodes

Figure 8.20 Electrode configuration: four electrodes are designated as forcers (FX and FY) and four are designated as pick-off (PX and PY). Both the forcer and pick-off channels have differential pairs (i.e. FX+ and FX−).

by the wafer to resonator alignment accuracy. Finally, the metal traces for the electrodes can be defined on the same material used for the resonator (i.e. fused silica), providing uniform coefficient of thermal expansion between the electrode die and the resonator.

Another important parameter to consider is the ratio of out-of-plane motion to in-plane motion, which indicates the transduction efficiency of the out-of-plane electrodes. Finite element modeling using Comsol Multiphysics package shows that, for mushroom type geometries the ratio of out-of-plane motion to in-plane motion is close to 1 : 1, leading to very efficient out-of-plane transduction, Figure 8.21.

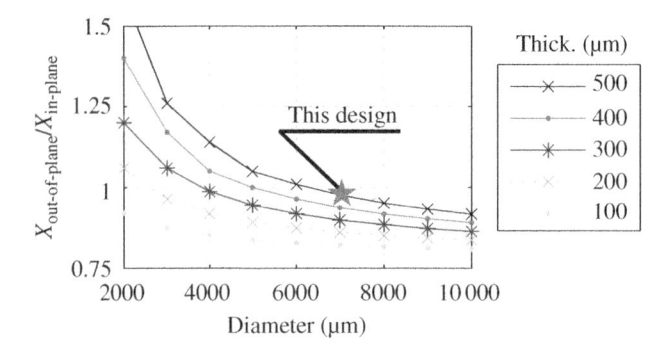

Figure 8.21 Out-of-plane to in-plane displacement ratio for mushroom-shaped resonators. Due to the 3-D nature of the resonator, the ratio is close to 1 : 1. Star marks the design presented in this chapter.

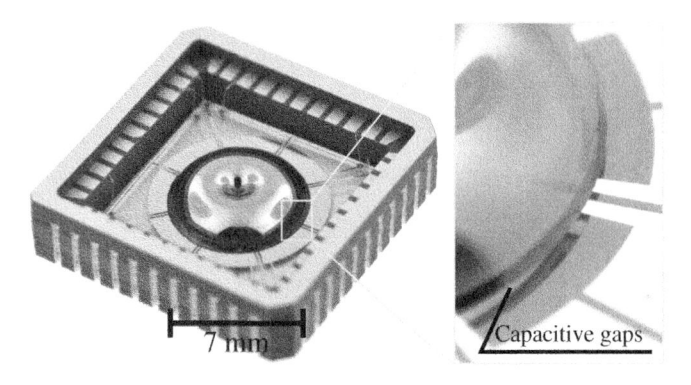

Figure 8.22 A packaged one million Q-factor fused silica wineglass structure with integrated electrodes (left), close-up of capacitive gaps (right).

8.7 Fabrication

Fabrication process utilizes two wafers: a wineglass shell wafer and an electrode wafer. Fabrication process starts with LPCVD polysilicon deposition on 1 mm thick fused silica wafers of up to 2 μm thickness, Figure 8.22. The polysilicon mask is later patterned lithographically and is used to etch cavities into fused silica wafers down to ~300 μm in depth. Once the etch is complete, the poly-silicon mask is removed using a 45% KOH bath. The next step of the fabrication process is plasma-assisted fusion bonding of a 500 μm thick fused silica device layer (Corning 7980) [13], Figure 8.23a. The plasma-assisted fusion bonding process for bonding fused silica wafer pairs can be summarized as follows:

1) Cleaning of the wafer pair using solvent and RCA clean,

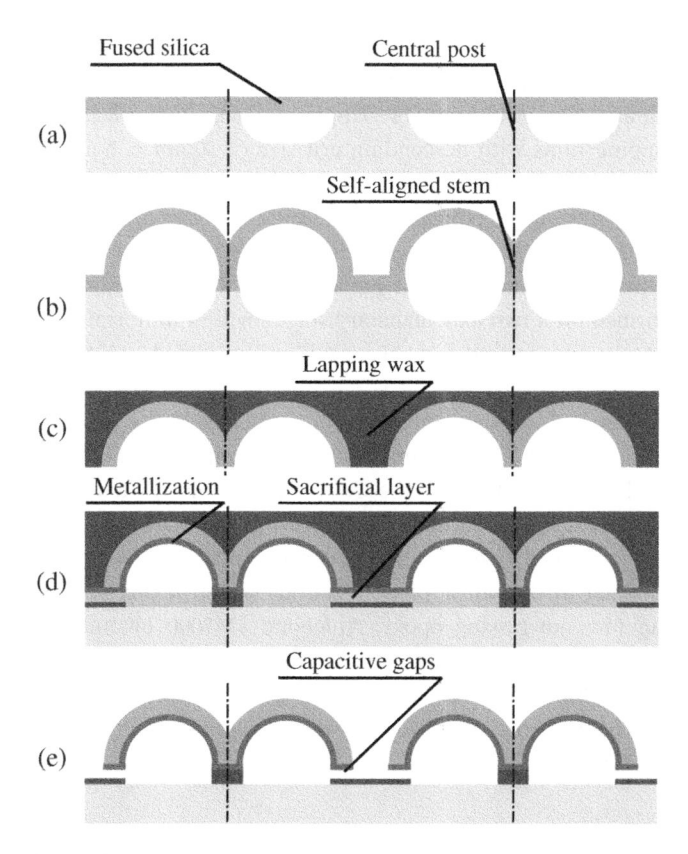

Figure 8.23 Wafer-level fabrication process for fused silica micro-wineglass structures: (a) plasma bonding of device layer to substrate with pre-etched cavities, (b) micro-glassblowing at 1700°C, (c) removal of the substrate via back-lapping, (d) bonding the wineglass wafer to electrode wafer, and (e) removal of the sacrificial layer to form capacitive gaps.

2) Plasma activation using oxygen plasma (50 W power for 2 min, 24 sccm O_2 flow),
3) DI water rinse followed by N_2 dry,
4) Contacting of the activated surfaces,
5) Room temperature anneal for >48 h,
6) Curing the wafer stack at 400°C for 6 h.

The bond creates a seamless hermetic seal around the etched cavities without using any intermediate material. The glassblowing is performed at 1700 for 2 min and rapidly cooled to room temperature, Figure 8.23b. During glassblowing, the device layer at the central post merges to create a solid, self-aligned stem structure, critical for high-Q operation. Shells are released by back-lapping the wafer stack to release using an Allied Multiprep 12″ lapping system, Figure 8.23c. A series of diamond lapping films with descending grit size of 30 µm ⇒ 6 µm ⇒ 3 µm ⇒ 1 µm ⇒ 0.5 µm ⇒ 0.1 µm are used for lapping, followed by final polish using 50 nm colloidal suspension and polishing cloth. Interior surface of the wineglasses is metallized with 50 nm thick sputtered Iridium. Only the interior surface is metallized to minimize the influence of the metal film on the shell resonator. Metallization is performed on a two axis planetary stage for film uniformity, in which the wafer is continuously rotated along the z-axis and oscillated ±15° along the x-axis.

Fused silica out-of-plane electrode structures are fabricated on a separate wafer by blanket coating with Cr/Au (100 nm/500 nm), spinning a thin layer of photo-resist sacrificial layer and patterning the Cr/Au features using etch-back. In this process, the photoresist is used both to pattern the electrodes and as a sacrificial layer to create the capacitive gaps. Subsequently, lapped and metallized wineglass wafer is bonded to the out-of-plane electrode wafer at the stem of each wineglass using low out-gassing epoxy, Ablebond JM7000 or Indium, Figure 8.23d. Once the bonding is complete, the sacrificial layer is removed to release the inverted wineglass structures around their perimeter, Figure 8.23e, creating capacitive gaps between the metallized inverted wineglass structures and the Cr/Au electrodes, Figure 8.24.

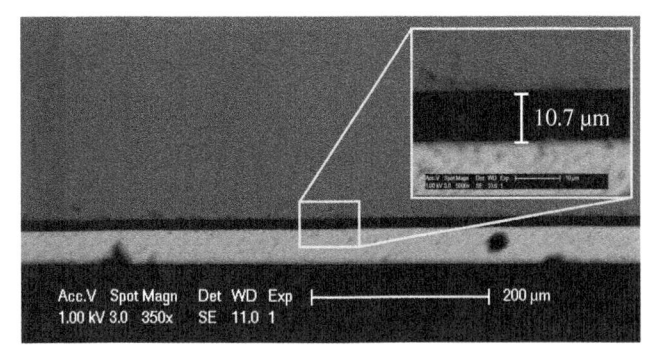

Figure 8.24 Uniform 10 µm capacitive gaps have been demonstrated on 7 mm shell structures, resulting in over 9 pF total active capacitance on the device.

8.8 Experimental Characterization

Frequency sweep using out-of-plane electrodes with ~30 μm capacitive gaps revealed a Q-factor of 1.14 million and a frequency split of 14 Hz at a center frequency of 105 kHz ($\Delta f/f$ = 132 ppm), Table 8.2 and Figure 8.25. Frequency sweep using the same set of forcer and pick-off electrodes showed an amplitude difference of ~30 dB between the two modes, indicative of misalignment between the electrodes and the principle axis of elasticity and/or quadrature coupling between the two modes. A separate ring-down experiment was performed where the device was excited with a narrow bandwidth swept sine-wave impulse and resonator output during free vibration was recorded. Ring-down experiment demonstrated a time constant of 3.18 s and a Q-factor of 1.05 million, confirming the frequency sweeps, Figure 8.26. In order to observe the effect of viscous damping on the overall Q-factor, the frequency sweep was repeated at different pressure levels. A Q-factor of 1 million was obtained below <20 μTorr, Figure 8.27. No further improvement in Q-factor was observed at 15 μTorr. Subsequently, capacitive gaps as low as 10 μm have been demonstrated on other 7 mm shells, resulting in over 9 pF of total active capacitance within the device (the total dC/dx of 970 nF/m), Figure 8.24.

Even though a fairly low frequency split of 14 Hz (132 ppm) was measured, this number is larger than previously reported 0.16 Hz (5.67 ppm) on dry etched borosilicate glass wineglass structures [145]. The increase in frequency split is attributed to the introduction of back-lapping process, which can induce asymmetry in the structure due to edge roughness. High performance degenerate mode gyroscope operation requires an even higher degree of symmetry in order to

Table 8.2 Summary of device parameters for a 7 mm fused silica wineglass resonator.

Shell diameter	7 mm
Shell thickness	~500 μm
Effective mass	8 μg
Quality factor	>1 million
Center frequency	105 kHz
Frequency split (Δf)	14 Hz

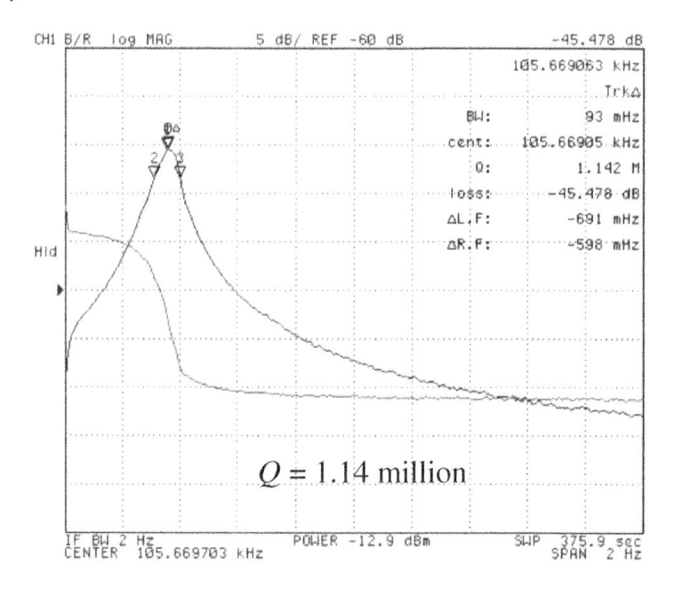

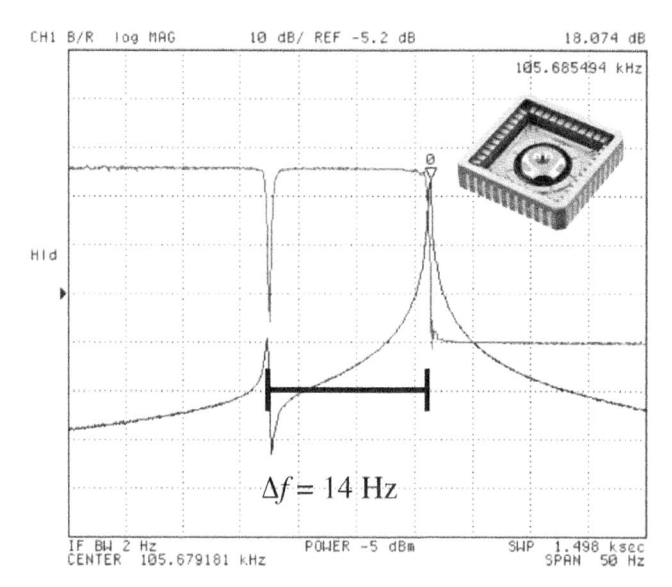

Figure 8.25 Frequency sweep revealed a Q-factor of 1.14 million and as fabricated frequency split (Δf) of 14 Hz at 105 kHz center frequency. The chamber pressure was 19 μTorr during the frequency sweep.

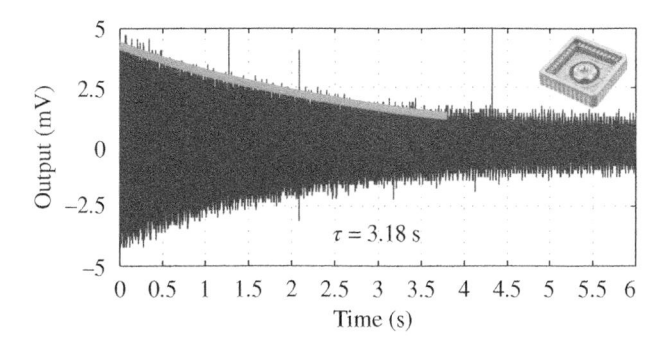

Figure 8.26 Ring-down experiment at 19 μTorr shows $\tau = 3.18$ s, giving 1.05 million Q-factor at 105 kHz, confirming the frequency sweep.

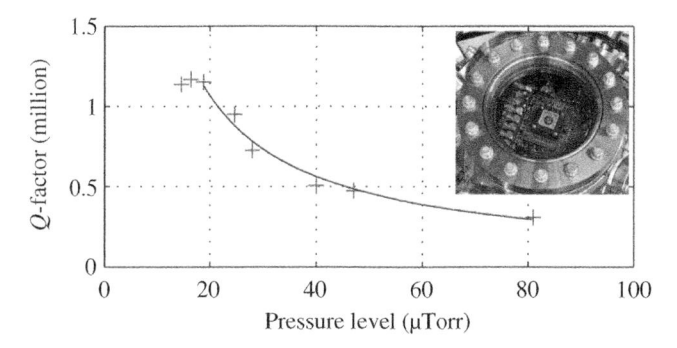

Figure 8.27 Q-factor versus pressure level experiment. Q-factors above 1 million were obtained below 20 μTorr.

leverage the high Q-factors seen on 3-D fused silica wineglass resonators. Further improvement in the back-lapping process and addition of postlapping surface treatment steps might help improve the structural symmetry further.

9

Conclusions and Future Trends

With the introduction of novel whole-angle gyroscope architectures and 3-D micro-machining techniques, batch fabrication of micro-scale whole-angle gyroscopes is becoming a reality. These new gyroscope architectures would not only help to improve our understanding of fundamental error sources affecting gyroscopes at micro-scale, but may also provide increased transduction efficiency, better rejection of environmental effects, and improved robustness at a smaller size. Meanwhile, new 3-D micro-machining techniques enable fabrication of high aspect ratio resonators out of low internal loss materials, such as fused silica and polycrystalline diamond, with greater precision than ever before. Low internal loss materials along with high structural symmetry and smoothness of these new 3-D resonators may enable fabrication of a new class of dynamic MEMS devices on the wafer surface at a significantly lower cost than their precision-machined, macro-scale counterparts.

Despite these recent advances, challenges related to scalability, robustness, and integration need to be addressed before micro-scale whole-angle gyroscopes can be realized at a commercial scale. Increased complexity associated with whole-angle gyroscopes create opportunities for novel solutions in fabrication technology, control systems, and packaging.

9.1 Mechanical Trimming of Structural Imperfections

The ability to achieve a high degree of structural symmetry on batch fabricated devices in a repeatable manner is essential for scalability of micro-scale whole-angle gyroscopes. Aside from optimization of fabrication processes for improved precision and repeatability, another approach for achieving desired structural symmetry is postfabrication mechanical trimming of gyroscopes.

Mechanical trimming of whole-angle gyroscopes starts with identifying the magnitude and direction of structural imperfections, principle axis of elasticity, and frequency asymmetry, via characterization of the resonator element. This is followed by the use of additive or subtractive processes to create a permanent change in modal mass or stiffness of the primary modes of the gyroscope. Compared to nonpermanent trimming approaches, such as electrostatic tuning, mechanical trimming can provide a greater degree of stability and a larger tuning range since this approach does not depend on an external factors such as a voltage source.

For example, a method for calculating mass matrix perturbations to improve linear vibration rejection of a Disk Resonator Gyroscope (DRG) was developed in [53]. The method was demonstrated on a macro-scale DRG prototype by placing NdFeB magnets on the spokes of the gyroscope to modify the mass distribution within the resonator element.

A model-based approach for reducing frequency asymmetries in a micro-scale DRG was described in [52]. Gold balls with approximately 20μg mass were used for "rough tuning." Due to finite minimum mass of gold balls, this was followed by "fine tuning" using silver ink. Frequency split of the gyroscope was successfully reduced from 14.1 Hz to < 0.1 Hz using this approach.

A wafer-scale etch process for mechanical trimming of MEMS gyroscopes was demonstrated in [54]. The approach relies on laser ablation of a protective, conformal layer covering the silicon resonator element, followed by Deep Reactive Ion Etching for mass removal at laser ablated locations. This approach was used to demonstrate frequency trimming of DRGs down to $\Delta f < 0.1$ Hz at a wafer-scale.

Postfabrication compensation of structural imperfections via mechanical trimming may help to reduce over-reliance on tight wafer-level fabrication tolerances and help to achieve the desired level of repeatability for commercialization.

9.2 Self-calibration

Self-calibration refers to in situ re-calibration techniques for correcting scale-factor and bias errors in gyroscopes. Not to be confused with sensor fusion, self-calibration typically refers to techniques that rely on information internal to the sensor and does not rely on information from external sensors such as accelerometers, magnetometers, GPS, etc. "Self-calibration" can be employed in a variety of ways depending on the frequency of calibration:

- Postfabrication self-calibration refers to one time re-calibration of the sensor after the device has been fabricated. This is typically done after mounting of the gyroscope onto the host system to correct for errors induced by mounting stresses, surface mount of components, and thermal re-flow process.
- Opportunistic self-calibration refers to self-calibration of the sensor based on an event. Opportunistic self-calibration takes place during a period of time where rate input is zero or sensor input is not needed.

- Continuous self-calibration refers to periodic re-calibration of the sensor during run-time.

As described in Chapter 3, one of the advantages of the whole-angle mechanization is the capability to do Virtual Carouseling, which can be used for system identification. Virtual Carouseling has the advantage of making error sources like aniso-elasticity, aniso-damping, electrode misalignments, x–y forcer, and pick-off gain unbalance observable. Characterization of these parameters through Virtual Carouseling is the first step on the path to continuous, run-time self-calibration of scale factor and bias errors.

9.3 Integration and Packaging

In addition to a high precision mechanical element, miniaturization is critical for realization of whole-angle gyroscopes in real-life applications. Miniaturization of the sensor assembly requires not only a small mechanical element, but also a close integration with electronics along with efficient packaging and a strong hermetic seal. 3-D nature of wineglass gyroscopes makes miniaturization even more challenging due to the convex sensor geometry, which creates opportunities for novel integration and packaging strategies.

For example, micro-glassblown wineglass resonators with out-of-plane electrode structures were reported in [135, 148]. Due to simplicity of the out-of-plane electrode architecture, the process is amenable to integration with CMOS processes. This can be achieved by replacing the fused silica out-of-plane electrode wafer with a CMOS wafer, so that the metal electrode structures that were previously defined on the fused silica wafer, can instead be defined on the top layer of the CMOS wafer. In addition to reduction in overall system size, integration of CMOS into the fabrication process would permit placement of front-end electronics close to electrode structures, which has potential advantages in terms of reduction of parasitic capacitances and improving signal-to-noise ratio.

Another example of close integration is a process flow for co-fabrication of wafer-level packages around micro-glassblown wineglass gyroscopes, described in [149]. The process relies on simultaneously co-fabricating micro-glassblowing cap and device layers to create a micro-wineglass shell resonator inside a micro-glassblown cavity, Figure 9.1. In addition to a potential reduction in size and cost, such a process would permit co-fabrication of the mechanical element and the cap layer from the same material, minimizing thermal drifts associated with the Coefficient of Thermal Expansion (CTE) mismatch between the two layers.

Leveraging the process described in [149], Shell-in-Shell or Dual-shell architectures, have been already developed and reported in [150–152]. Devices with dual-shell configuration are demonstrating the frequency ranges from 5kHz to

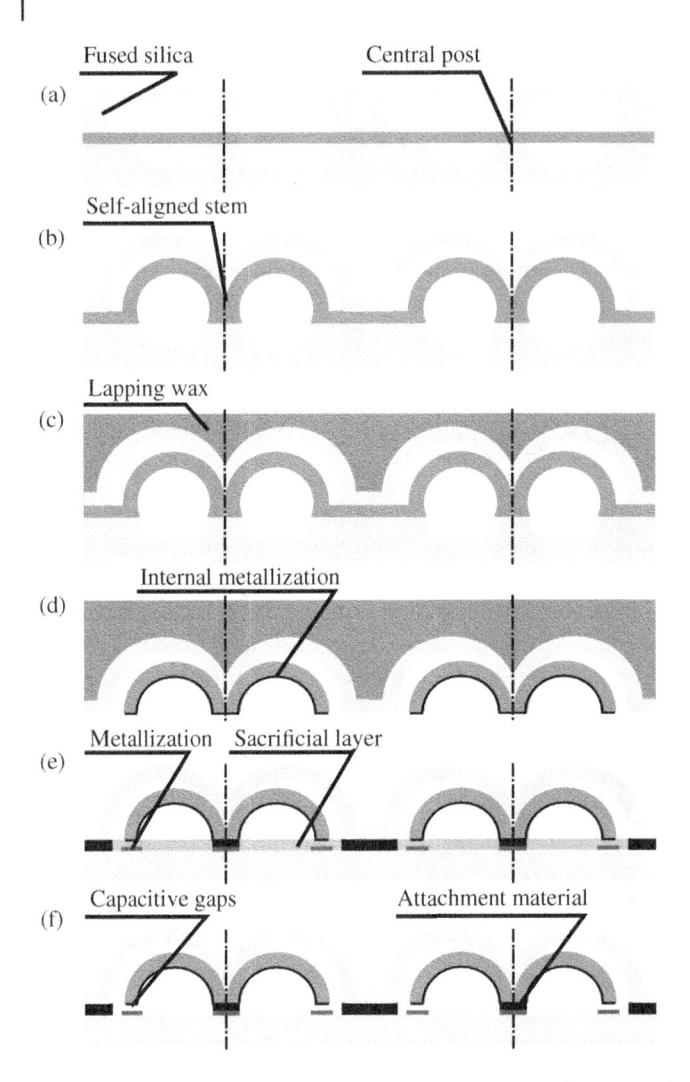

Figure 9.1 Fabrication process consists of (a) bonding of pre-etched cap and substrate wafers to the device layer, (b) micro-glassblowing the wafer stack, (c) embedding the stack in support material, (d) removal of the substrate wafer via lapping, (e) bonding the device layer with cap to the interposer/electrode wafer, and (f) removing support and sacrificial material to create capacitive gaps.

several MHz, the amplitude ringdown time of more than 100 seconds, the compensated frequency mismtach <300 mHz, the scale factor of 5.5 mV/deg/s, the Allan deviation of zero-rate output on initial prototypes revealed a bias instability on the level of 0.4 deg/hr and ARW of 0.06 deg/rt-hr [153].

References

1 A. M. Shkel. Type I and Type II Micromachined Vibratory Gyroscopes. In IEEE/ION Position Location and Navigation Symposium (PLANS), San Diego, CA, USA, 2006.
2 D. Senkal. *Micro-Glassblowing Paradigm for Realization of Rate Integrating Gyroscopes*. PhD thesis, University of California, Irvine, 2015.
3 IEEE Standard Specification Format Guide and Test Procedure for Single-Axis Interferometric Fiber Optic Gyros - IEEE Std 952. *Technical report*, IEEE Standards Board, 1997.
4 D. D. Lynch. Vibratory Gyro Analysis by the Method of Averaging. In Saintt Petersburg Conference on Gyroscopic Technology and Navigation, Saint Petersburg, Russia, 1995.
5 M. Weinberg, R. Candler, S. Chandorkar, J. Varsanik, T. Kenny, and A. Duwel. Energy Loss in MEMS Resonators and the Impact on Inertial and RF Devices. In Solid-State Sensors, Actuators and Microsystems Conference (TRANSDUCERS), Denver, CO, USA, 2009.
6 V. Kaajakari. Practical MEMS. Small Gear Publishing, Las Vegas, NV, USA, 2009.
7 R. N. Candler, M. a. Hopcroft, B. Kim, W.-T. Park, R. Melamud, M. Agarwal, G. Yama, A. Partridge, M. Lutz, and T. W. Kenny. Long-Term and Accelerated Life Testing of a Novel Single-Wafer Vacuum Encapsulation for MEMS Resonators. *IEEE/ASME Journal of Microelectromechanical Systems*, 15(6):1446–1456, 2006.
8 M. H. Asadian, S. Askari, and A. M. Shkel. An Ultrahigh Vacuum Packaging Process Demonstrating Over 2 Million Q-Factor in MEMS Vibratory Gyroscopes. *IEEE Sensors Letters*, 1(6):1–4, 2017.
9 Z. Hao and F. Ayazi. Thermoelastic Damping in Flexural-Mode Ring Gyroscopes. In International Mechanical Engineering Congress and Exposition, Orlando, FL, USA, 2005.
10 C. Zener. Internal Friction in Solids I. Theory of Internal Friction in Reeds. *Physical Review*, 52:230–234, 1937.

11 C. Zener. Internal Friction in Solids II. General Theory of Thermoelastic Internal Friction. *Physical Review*, 53:90–99, 1938.

12 B. S. Lunin. Physical and Chemical Bases for the Development of Hemispherical Resonators for Solid-State Gyroscopes. Technical report, Moscow Aviation Institute, Moscow, 2005.

13 D. Senkal, M. J. Ahamed, A. A. Trusov, and A. M. Shkel. High Temperature Micro-Glassblowing Process Demonstrated on Fused Quartz and ULE TSG. *Sensors and Actuators A: Physical*, 201:525–531, 2012.

14 S. Chandorkar, M. Agarwal, R. Melamud, R. N. Candler, K. E. Goodson, and T. W. Kenny. Limits of Quality Factor in Bulk-Mode Micromechanical Resonators. In IEEE International Conference on Micro Electro Mechanical Systems (MEMS), Tucson, AZ, USA, 2008.

15 B. Friedland and M. Hutton. Theory and Error Analysis of Vibrating-Member Gyroscope. *IEEE Transactions on Automatic Control*, 23(4):545–556, 1978.

16 C. C. Painter and A. M. Shkel. Active Structural Error Suppression in MEMS Vibratory Rate Integrating Gyroscopes. *IEEE Sensors Journal*, 3(5):595–606, 2003.

17 D. M. Rozelle. The Hemispherical Resonator Gyro: From Wineglass to the Planets. In AAS/AIAA Space Flight Mechanics Meeting, Savannah, GA, USA, 2009.

18 T. Niu and M. Palaniapan. A Low Phase Noise 10 MHz Micromechanical Lamé-Mode Bulk Oscillator Operating in Nonlinear Region. In IEEE Frequency Control Symposium (FCS), Newport Beach, CA, USA, 2010.

19 Y. Okada and Y. Tokumaru. Precise Determination of Lattice Parameter and Thermal Expansion Coefficient of Silicon Between 300 and 1500 K. *Journal of Applied Physics*, 56(2):314, 1984.

20 T. Ray, H. Zhu, and I. Elango. Characterizaton of KMPR 1025 as a Masking Layer for Deep Reactive Ion Etching of Fused Silica. In IEEE International Conference on Micro Electro Mechanical Systems (MEMS), Cancun, Mexico, 2011.

21 K. Kolari, V. Saarela, and S. Franssila. Deep Plasma Etching of Glass for Fluidic Devices with Different Mask Materials. *Journal of Micromechanics and Microengineering*, 18(6):1–6, 2008.

22 D. D. Lynch. Coriolis Vibratory Gyros. In Symposium Gyro Technology, Stuttgart, Germany, 1998.

23 K. M. Harish, B. J. Gallacher, J. S. Burdess, and J. A. Neasham. Experimental Investigation of Parametric and Externally Forced Motion in Resonant MEMS Sensors. *Journal of Micromechanics and Microengineering*, 19(1):015021, 2009.

24 L. A. Oropeza-Ramos, C. B. Burgner, and K. L. Turner. Robust Micro-rate Sensor Actuated by Parametric Resonance. *Sensors and Actuators A: Physical*, 152(1):80–87, 2009.

25 A. M. Shkel and R. T. Howe. Micro-Machined Angle-Measuring Gyroscope. *US Patent 6,481,285*, 2002.

26 C. C. Painter and A. M. Shkel. Experimental Evaluation of a Control System for an Absolute Angle Measuring Micromachined Gyroscope. In IEEE Sensors Conference, Irvine, CA, USA, 2005.

27 D. Wang, A. Efimovskaya, and A. M. Shkel. Amplitude Amplified Dual-Mass Gyroscope: Design Architecture and Noise Mitigation Strategies. In IEEE International Symposium on Inertial Sensors and Systems, Naples, FL, USA, 2019.

28 J. Giner, D. Maeda, K. Ono, A. M. Shkel, and T. Sekiguchi. MEMS Gyroscope with Concentrated Springs Suspensions Demonstrating Single Digit Frequency Split and Temperature Robustness. *IEEE/ASME Journal of Microelectromechanical Systems*, 28(1):25–35, 2019.

29 A. A. Trusov, A. R. Schofield, and A. M. Shkel. Micromachined Tuning Fork Gyroscopes with Ultra-High Sensitivity and Shock Rejection. *US Patent 8,322,213*, 2012.

30 I. P. Prikhodko, A. A. Trusov, and A. M. Shkel. Compensation of Drifts in High-Q MEMS Gyroscopes Using Temperature Self-Sensing. *Sensors and Actuators A: Physical*, 201:517–524, 2013.

31 I. P. Prikhodko, S. A. Zotov, A. A. Trusov, and A. M. Shkel. Foucault Pendulum on a Chip: Rate Integrating Silicon MEMS Gyroscope. *Sensors and Actuators, A: Physical*, 177:67–78, 2012.

32 A. A. Trusov, D. M. Rozelle, G. Atikyan, S. A. Zotov, B. R. Simon, A. M. Shkel, and A. D. Meyer. Non-Axisymmetric Coriolis Vibratory Gyroscope with Whole-Angle, Force Rebalance, and Self-Calibration. In Solid-State Sensors, Actuators, and Microsystems Workshop (Hilton Head), Hilton Head Island, SC, USA, 2014.

33 S. Askari, M. H. Asadian, and A. M. Shkel. High quality factor MEMS gyroscope with whole-angle mode of operation. In IEEE International Symposium on Inertial Sensors and Systems, Moltrasio, Italy, 2018.

34 S. Askari, M. H. Asadian, and A. M. Shkel. Retrospective Correction of Angular Gain by Virtual Carouseling in MEMS Gyroscopes. In IEEE International Symposium on Inertial Sensors and Systems, Naples, FL, USA, 2019.

35 C. Guo, E. Tatar, and G. K. Fedder. Large-Displacement Parametric Resonance Using a Shaped Comb Drive. In IEEE International Conference on Micro Electro Mechanical Systems (MEMS), Taipei, Taiwan, 2013.

36 E. Tatar, T. Mukherjee, and G. K. Fedder. Simulation of Stress Effects on Mode-Matched MEMS Gyroscope Bias and Scale Factor. In IEEE/ION Position Location and Navigation Symposium (PLANS), Monterey, CA, USA, 2014.

37 E. Tatar, T. Mukherjee, and G. K. Fedder. On-Chip Characterization of Stress Effects on Gyroscope Zero Rate Output and Scale Factor. In IEEE International

Conference on Micro Electro Mechanical Systems (MEMS), Estoril, Portugal, 2015.

38 M. P. Varnham, D. Hodgins, T. S. Norris, and H. D. Thomas. Vibrating Planar Gyro. *US Patent 5,226,321*, 1993.

39 B. Gallacher, J. Hedley, J. Burdess, A. Harris, A. Rickard, and D. King. Electrostatic Correction of Structural Imperfections Present in a Microring Gyroscope. *IEEE/ASME Journal of Microelectromechanical Systems*, 14(2):221–234, 2005.

40 B. Gallacher. Principles of a Micro-Rate Integrating Ring Gyroscope. *IEEE Transaction on Aerospace and Electronic Systems*, 48(1):658–672, 2012.

41 S. R. Bowles, B. J. Gallacher, Z. X. Hu, C. Gregory, and K. Townsend. Control Scheme for a Rate Integrating MEMS Gyroscope. In IEEE International Symposium on Inertial Sensors and Systems, Laguna Beach, CA, USA, 2014.

42 F. Ayazi and K. Najafi. Design and Fabrication of High-Performance Polysilicon Vibrating Ring Gyroscope. In IEEE International Conference on Micro Electro Mechanical Systems (MEMS), Heidelberg, Germany, 1998.

43 F. Ayazi, H. Chen, F. Kocer, H. Guohong, and K. Najafi. A High Aspect-Ratio Polysilicon Vibrating Ring Gyroscope. In Solid-State Sensors, Actuators, and Microsystems Workshop (Hilton Head), Hilton Head Island, SC, USA, 2000.

44 F. Ayazi and K. Najafi. A HARPSS Polysilicon Vibrating Ring Gyroscope. *IEEE/ASME Journal of Microelectromechanical Systems*, 10(2):169–179, 2001.

45 F. Ayazi and K. Najafi. High Aspect-Ratio Polysilicon Micromachining Technology. *Sensors and Actuators A: Physical*, 87(1–2):46–51, 2000.

46 F. Ayazi and K. Najafi. High Aspect-Ratio Combined Poly and Single-Crystal Silicon (HARPSS) MEMS Technology. *IEEE/ASME Journal of Microelectromechanical Systems*, 9(3):288–294, 2000.

47 A. D. Challoner and K. V. Shcheglov. Isolated Resonator Gyroscope with a Drive and Sense Plate. *US Patent 7,093,486*, 2006.

48 A. D. Challoner, H. H. Ge, and J. Y. Liu. Boeing Disc Resonator Gyroscope. In IEEE/ION Position Location and Navigation Symposium (PLANS), Savannah, GA, USA, 2014.

49 R. Kubena and D. Chang. Disc Resonator Gyroscopes. *US Patent 7,581,443*, 2009.

50 D. Kim and R. T. M'Closkey. Noise Analysis of Closed–Loop Vibratory Rate Gyros. In American Control Conference (ACC), Montreal, QC, Canada, 2012.

51 D. Kim and R. T. M'Closkey. Spectral Analysis of Vibratory Gyro Noise. *IEEE Sensors Journal*, 13(11):4361–4374, 2013.

52 D. Schwartz, D. Kim, and R. M. Closkey. A Model-Based Approach to Multi-Modal Mass Tuning of a Micro-Scale Resonator. In IEEE American Control Conference (ACC), Montreal, QC, Canada, 2012.

53 D. Schwartz and R. T. M'Closkey. Decoupling of a Disk Resonator From Linear Acceleration Via Mass Matrix Perturbation. *Journal of Dynamic Systems, Measurement, and Control*, 134(2):021005, 2012.

54 D. Kim, A. Behbahani, R. T. M'Closkey, P. Stupar, and J. Denatale. Wafer-Scale Etch Process for Precision Frequency Tuning of MEMS Gyros. In IEEE International Symposium on Inertial Sensors and Systems, Hapuna Beach, HI, USA, 2015.

55 E. Ng, H. Lee, C. Ahn, R. Melamud, and T. W. Kenny. Stability Measurements of Silicon MEMS Resonant Thermometers. In IEEE Sensors Conference, Limerick, Ireland, 2011.

56 S. Nitzan, C. H. Ahn, T.-H. Su, M. Li, E. J. Ng, S. Wang, Z. M. Yang, G. O'Brien, B. E. Bose, T. W. Kenny, and D. A. Horsley. Epitaxially-Encapsulated Polysilicon Disk Resonator Gyroscope. In IEEE International Conference on Micro Electro Mechanical Systems (MEMS), Taipei, Taiwan, 2013.

57 C. H. Ahn, E. J. Ng, V. A. Hong, Y. Yang, B. J. Lee, M. W. Ward, and T. W. Kenny. Geometric Compensation of (100) Single Crystal Silicon Disk Resonating Gyroscope for Mode-Matching. In Solid-State Sensors, Actuators and Microsystems Conference (TRANSDUCERS), Barcelona, Spain, 2013.

58 S. Nitzan, T. H. Su, C. Ahn, E. Ng, V. Hong, Y. Yang, T. W. Kenny, and D. A. Horsley. Impact of Gyroscope Operation Above the Critical Bifurcation Threshold on Scale Factor and Bias Instability. In IEEE International Conference on Micro Electro Mechanical Systems (MEMS), San Francisco, CA, USA, 2014.

59 P. Taheri-Tehrani, O. Izyumin, I. Izyumin, C. H. Ahn, E. J. Ng, V. A. Hong, Y. Yang, T. W. Kenny, B. E. Boser, and D. A. Horsley. Disk Resonator Gyroscope with Whole-Angle Mode Operation. In IEEE International Symposium on Inertial Sensors and Systems, Hapuna Beach, HI, USA, 2015.

60 J. Cho, J. Gregory, and K. Najafi. Single-Crystal-Silicon Vibratory Cylinderical Rate Integrating Gyroscope (CING). In Solid-State Sensors, Actuators and Microsystems Conference (TRANSDUCERS), Beijing, China, 2011.

61 J. Y. Cho. *High-Performance Micromachined Vibratory Rate and Rate-integrating Gyroscopes*. PhD thesis, University of Michigan, 2012.

62 J. Gregory, J. Cho, and K. Najafi. MEMS Rate and Rate-Integrating Gyroscope Control with Commercial Software Defined Radio Hardware. In Solid-State Sensors, Actuators and Microsystems Conference (TRANSDUCERS), Beijing, China, 2011.

63 J. Gregory, J. Cho, and K. Najafi. Novel Mismatch Compensation Methods for Rate-Integrating Gyroscopes. In IEEE/ION Position Location and Navigation Symposium (PLANS), Myrtle Beach, SC, USA, 2012.

64 J. Gregory, J. Cho, and K. Najafi. Characterization and Control of a High-Q MEMS Inertial Sensor Using Low-Cost Hardware. In IEEE/ION Position Location and Navigation Symposium (PLANS), Myrtle Beach, SC, USA, 2012.

65 J. Cho, J. Gregory, and K. Najafi. High-Q, 3 kHz Single-Crystal-Silicon Cylindrical Rate-Integrating Gyro (CING). In IEEE International Conference on Micro Electro Mechanical Systems (MEMS), Paris, France, 2012.

66 H. Johari and F. Ayazi. Capacitive Bulk Acoustic Wave Silicon Disk Gyroscopes. In IEEE International Electron Devices Meeting, San Francisco, CA, USA, 2006.

67 H. Johari. *Micromachined Capacitive Silicon Bulk Acoustic Wave Gyroscopes*. PhD thesis, Georgia Institute of Technology, 2008.

68 H. Johari and F. Ayazi. High-Frequency Capacitive Disk Gyroscopes in (100) and (111) Silicon. In IEEE International Conference on Micro Electro Mechanical Systems (MEMS), Hyogo, Japan, 2007.

69 F. Ayazi and H. Johari. Capacitive Bulk Acoustic Wave Disk Gyroscopes. *US Patent 7,543,496*, 2009.

70 J. Seeger, M. Lim, and S. Nasiri. Development of High-Performance, High-Volume Consumer MEMS Gyroscopes. Technical report, TDK Invensense Inc., Sunnyvale, CA, USA, 2010.

71 Z. Hao, S. Pourkamali, and F. Ayazi. VHF Single-Crystal Silicon Elliptic Bulk-Mode Capacitive Disk Resonators-Part I: Design and Modeling. *IEEE/ASME Journal of Microelectromechanical Systems*, 13(6):1043–1053, 2004.

72 D. Senkal, S. Askari, M. J. Ahamed, E. J. Ng, V. Hong, Y. Yang, C. H. Ahn, T. W. Kenny, and A. M. Shkel. 100k Q-Factor Toroidal Ring Gyroscope Implemented in Wafer-Level Epitaxial Silicon Encapsulation Process. In IEEE International Conference on Micro Electro Mechanical Systems (MEMS), Taipei, Taiwan, 2014.

73 A. Efimovskaya, D. Wang, Y. W. Lin, and A. M. Shkel. On Ordering Of Fundamental Wineglass Modes in Toroidal Ring Gyroscope. In Proceedings of IEEE Sensors, Orlando, FL, USA, 2017.

74 D. Senkal, E. J. Ng, V. Hong, Y. Yang, C. H. Ahn, T. W. Kenny, and A. M. Shkel. Parametric Drive of a Toroidal MEMS Rate Integrating Gyroscope Demonstrating 20 ppm Scale Factor Stability. In IEEE International Conference on Micro Electro Mechanical Systems (MEMS), Estoril, Portugal, 2015.

75 D. Senkal, A. Efimovskaya, and A. M. Shkel. Minimal Realization of Dynamically Balanced Lumped Mass WA Gyroscope: Dual Foucault Pendulum. In IEEE International Symposium on Inertial Sensors and Systems, Hapuna Beach, HI, USA, 2015.

76 D. Senkal, A. Efimovskaya, and A. M. Shkel. Dual Foucault Pendulum Gyroscope. In Solid-State Sensors, Actuators and Microsystems Workshop (TRANSDUCERS), Anchorage, AK, USA, 2015.

77 J. Bernstein, S. Cho, A. T. King, P. Kourepenis, P. Maciel, and M. Wienberg. A Micromachined Comb-Drive Tuning Fork Rate Gyroscope. In IEEE International Conference on Micro Electro Mechanical Systems (MEMS), Fort Lauderdale, FL, USA, 1993.

78 A. A. Trusov, G. Atikyan, D. M. Rozelle, A. D. Meyer, S. A. Zotov, B. R. Simon, and A. M. Shkel. Flat Is Not Dead: Current and Future Performance of Si-MEMS Quad Mass Gyro (QMG) System. In IEEE/ION Position Location and Navigation Symposium (PLANS), Savannah, GA, USA, 2014.

79 M. H. Asadian, S. Askari, I. B. Flader, Y. Chen, D. D. Gerrard, D. D. Shin, H.-K. Kwon, T. W. Kenny, and A. M. Shkel. High Quality Factor Mode Ordered Dual Foucault Pendulum Gyroscope. In IEEE Sensors Conference, New Delhi, India, 2018.

80 A. M. Shkel. DARPA-BAA-10-39: Microscale Rate Integrating Gyroscope (MRIG). DARPA Microsystems Technology Office, 2010.

81 E. J. J. Loper and D. D. Lynch. Sonic Vibrating Bell Gyro. *US Patent 4,157,041*, 1979.

82 L. Kumar, M. J. Foster, and T. A. Bittner. Vibratory Rotation Sensor with AC Forcing and Sensing Electronics. *US Patent 5,850,041*, 1998.

83 D. Lynch. Bell Gyro and Improved Means for Operating Same. *US Patent 3,656,354*, 1972.

84 A. Meyer and D. Rozelle. Milli-HRG Inertial Navigation System. In IEEE/ION Position Location and Navigation Symposium (PLANS), Myrtle Beach, SC, USA, 2012.

85 R. E. Stewart. Micro Hemispheric Resonator Gyro. *US Patent 8,109,145*, 2007.

86 G. Johnson. Vibratory Sensor with Self-Calibration and Low Noise Digital Conversion. *US Patent 6,189,382*, 2001.

87 A. Jeanroy. Gyroscopic Sensor. *US Patent 6,662,656*, 2000.

88 L. Rosellini and J. M. Caron. REGYS 20: A Promising HRG-Based IMU for Space Application. In International ESA Conference on Guidance, Navigation and Control Systems (GNC), Tralee, County Kerry, Ireland, 2008.

89 A. Jeanroy. Method for Calibrating a Scale Factor in an Axisymmetrical Vibrating Gyrometer. *US Patent 8,210,023*, 2008.

90 A. Jeanroy and P. Leger. Gyroscopic Sensor and Rotation Measurement Apparatus Constituting an Application Thereof. *US Patent 6,474,161*, 2002.

91 A. Renault. Method for Implementing a Resonator Under Electrostatic Forces. *US Patent 7,127,946*, 2006.

92 A. Renault and P. Vandebeuque. Hemispherical Resonator with Divided Shield Electrode. *US Patent 6,945,109*, 2005.

93 E. J. Eklund and A. M. Shkel. Glass Blowing on a Wafer Level. *IEEE/ASME Journal of Microelectromechanical Systems*, 16(2):232–239, 2007.

94 E. J. Eklund and A. M. Shkel. *Self-Inflated Micro-Glass Blowing*. US Patent 8,151,600, 2012.

95 E. Eklund, A. Shkel, S. Knappe, E. Donley, and J. Kitching. Glass-blown Spherical Microcells for Chip-Scale Atomic Devices. *Sensors and Actuators A: Physical*, 143(1):175–180, 2008.

96 M. A. Huff, A. D. Nikolich, and M. A. Schmidt. Design of Silicon Cavity Microstructures Formed by Silicon Wafer Bonding. *IEEE/ASME Journal of Microelectromechanical Systems*, 2(2):74–81, 1993.

97 S. A. Zotov, I. P. Prikhodko, A. A. Trusov, and A. M. Shkel. 3-D Micromachined Spherical Shell Resonators with Integrated Electromagnetic and Electrostatic Transducers. In Solid-State Sensors, Actuators, and Microsystems Workshop (Hilton Head), Hilton Head Island, SC, USA, 2010.

98 I. P. Prikhodko, S. A. Zotov, A. A. Trusov, and A. M. Shkel. Microscale Glass-Blown Three-Dimensional Spherical Shell Resonators. *IEEE/ASME Journal of Microelectromechanical Systems*, 20(3):691–701, 2011.

99 M. Kanik, P. Bordeenithikasem, D. Kim, N. Selden, A. Desai, R. M. Closkey, and J. Schroers. Metallic Glass Hemispherical Shell Resonators. *IEEE/ASME Journal of Microelectromechanical Systems*, 24(1):19–28, 2015.

100 M. Kanik, P. Bordeenithikasem, J. Schroers, D. Kim, and R. M'Closkey. Microscale Three-Dimensional Hemispherical Shell Resonators Fabricated from Metallic Glass. In IEEE International Symposium on Inertial Sensors and Systems, Laguna Beach, CA, USA, 2014.

101 B. Sarac, G. Kumar, T. Hodges, S. Ding, A. Desai, and J. Schroers. Three-Dimensional Shell Fabrication Using Blow Molding of Bulk Metallic Glass. *IEEE/ASME Journal of Microelectromechanical Systems*, 20(1):28–36, 2011.

102 J. Y. Cho, T. Nagourney, A. Darvishian, B. Shiari, J. Woo, and K. Najafi. Fused Silica Micro Birdbath Shell Resonators with 1.2 Million Q and 43 Second Decay Time Constant. In Solid-State Sensors, Actuators and Microsystems Workshop (Hilton Head), Hilton Head Island, SC, USA, 2014.

103 J. Y. Cho, J. Yan, J. A. Gregory, H. Eberhart, R. L. Peterson, and K. Najafi. High-Q Fused Silica Birdbath and Hemispherical 3-D Resonators Made by Blow Torch Molding. In IEEE International Conference on Micro Electro Mechanical Systems (MEMS), Taipei, Taiwan, 2013.

104 J. Y. Cho and K. Najafi. A High-Q All-Fused Silica Solid-Stem Wineglass Hemispherical Resonator Formed Using Micro Blow Torching and Welding. In IEEE International Conference on Micro Electro Mechanical Systems (MEMS), Estoril, Portugal, 2015.

105 K. Visvanathan, T. Li, and Y. B. Gianchandani. 3D-SOULE: A Fabrication Process for Large Scale Integration and Micromachining of Spherical Structures.

In IEEE International Conference on Micro Electro Mechanical Systems (MEMS), Cancun, Mexico, 2011.
106. L. C. Fegely, D. N. Hutchison, and S. A. Bhave. Isotropic Etching of 111 SCS for Wafer-Scale Manufacturing of Perfectly Hemispherical Silicon Molds. In Solid-State Sensors, *Actuators and Microsystems Conference (TRANSDUCERS)*, Beijing, China, 2011.
107. A. K. Bhat, L. C. Fegely, and S. A. Bhave. GOBLIT: A Giant Opto-Mechanical Bulk-Machined Light Transducer. In *Solid-State Sensors, Actuators and Microsystems Workshop (Hilton Head)*, Hilton Head Island, SC, USA, 2014.
108. P. Shao, L. D. Sorenson, X. Gao, and F. Ayazi. Wineglass-on-a-Chip. In Solid-State Sensors, Actuators, and Microsystems Workshop (Hilton Head), volume 7, Hilton Head Island, SC, USA, 2012.
109. P. Shao, V. Tavassoli, L. Chang-Shun, L. Sorenson, and F. Ayazi. Electrical Characterization of ALD-Coated Silicon Dioxide Micro-Hemispherical Shell Resonators. In IEEE International Conference on Micro Electro Mechanical Systems (MEMS), San Francisco, CA, USA, 2014.
110. N. Mehanathan, V. Tavassoli, P. Shao, L. Sorenson, and F. Ayazi. Invar-36 Micro Hemispherical Shell Resonators. In IEEE International Conference on Micro Electro Mechanical Systems (MEMS), San Francisco, CA, USA, 2014.
111. P. Shao, C. L. Mayberry, X. Gao, V. Tavassoli, and F. Ayazi. A Polysilicon Microhemispherical Resonating Gyroscope. *IEEE Journal of Microelectromechanical Systems*, 23(4):762–764, 2014.
112. M. L. Chan, J. Xie, P. Fonda, H. Najar, K. Yamazaki, L. Lin, and D. A. Horsley. Micromachined Polycrystalline Diamond Hemispherical Shell Resonators. In *Solid-State Sensors, Actuators, and Microsystems Workshop (Hilton Head)*, Hilton Head Island, SC, USA, 2012.
113. A. Heidari, M. Chan, H.-A. Yang, G. Jaramillo, P. Taheri-Tehrani, P. Fonda, H. Najar, K. Yamazaki, L. Lin, and D. A. Horsley. Micromachined Polycrystalline Diamond Hemispherical Shell Resonators. In *Solid-State Sensors, Actuators, and Microsystems Conference (TRANSDUCERS)*, Barcelona, Spain, 2013.
114. H. Najar, A. Heidari, M.-L. Chan, H.-A. Yang, L. Lin, D. G. Cahill, and D. A. Horsley. Microcrystalline Diamond Micromechanical Resonators with Quality Factor Limited by Thermoelastic Damping. *Applied Physics Letters*, 102(7):071901, 2013.
115. D. Saito, C. Yang, A. Heidari, H. Najar, L. Lin, and D. A. Horsley. Batch-Fabricated High Q-Factor Microcrystalline Diamond Cylindrical Resonator. In *IEEE International Conference on Micro Electro Mechanical Systems (MEMS)*, Estoril, Portugal, 2015.
116. R. Perahia, J. J. Lake, S. S. Iyer, D. J. Kirby, H. D. Nguyen, T. J. Boden, R. J. Joyce, L. X. Huang, L. D. Sorenson, and D. T. Chang. Electric Gradient Force Drive Mechanism for Novel Micro-Scale All-Dielectric Gyroscope. In IEEE

International Conference on Micro Electro Mechanical Systems (MEMS), San Francisco, CA, USA, 2014.

117 R. Perahia, H. D. Nguyen, L. X. Huang, T. J. Boden, J. J. Lake, D. J. Kirby, R. J. Joyce, L. D. Sorenson, and D. T. Chang. Novel Touch-Free Drive, Sense, and Tuning Mechanism for All-Dielectric Micro-Shell Gyroscope. In Solid-State Sensors, Actuators and Microsystems Workshop (Hilton Head), Hilton Head Island, SC, USA, 2014.

118 M. Rahman, Y. Xie, C. Mastrangelo, and H. Kim. 3-D Hemispherical Micro Glass-Shell Resonator with Integrated Electrostatic Excitation and Capacitive Detection Transducers. In IEEE International Conference on Micro Electro Mechanical Systems (MEMS), San Francisco, CA, USA, 2014.

119 J. J. Bernstein, M. G. Bancu, E. H. Cook, M. V. Chaparala, W. A. Teynor, and M. S. Weinberg. A MEMS Diamond Hemispherical Resonator. *Journal of Micromechanics and Microengineering*, 23(12):125007, 2013.

120 P. Pai, F. K. Chowdhury, C. H. Mastrangelo, and M. Tabib-Azar. MEMS-Based Hemispherical Resonator Gyroscopes. In *IEEE Sensors Conference*, Taipei, Taiwan, 2012.

121 P. Pai, F. K. Chowdhury, H. Pourzand, and M. Tabib-azar. Fabrication and Testing of Hemispherical MEMS Wineglass Resonators. In *IEEE International Conference on Micro Electro Mechanical Systems (MEMS)*, Taipei, Taiwan, 2013.

122 Y. Xie, H. C. Hsieh, P. Pai, H. Kim, M. Tabib-Azar, and C. H. Mastrangelo. Precision Curved Micro Hemispherical Resonator Shells Fabricated by Poached-Egg Micro-Molding. In *IEEE Sensors Conference*, Taipei, Taiwan, 2012.

123 B. R. Johnson, C. Boynton, E. Cabuz, S. Chang, K. Christ, S. Moore, J. Reinke, and K. Winegar. Toroidal Resonators with Small Frequency Mismatch for Rate Integrating Gyroscopes. In IEEE International Symposium on Inertial Sensors and Systems, Laguna Beach, CA, USA, 2014.

124 E. Hendarto, T. Li, and Y. B. Gianchandani. Investigation of Wine Glass Mode Resonance in 200-μm-Diameter Cenosphere-Derived Borosilicate Hemispherical Shells. *Journal of Micromechanics and Microengineering*, 23(5):055013, 2013.

125 L. D. Sorenson, P. Shao, and F. Ayazi. Effect of Thickness Anisotropy on Degenerate Modes in Oxide Micro-Hemispherical Shell Resonators. In IEEE International Conference on Micro Electro Mechanical Systems (MEMS), Taipei, Taiwan, 2013.

126 J. Y. Cho, J. K. Woo, J. Yan, J. L. Peterson, and K. Najafi. Fused-Silica Micro Birdbath Resonator Gyroscope (μ-BRG). *IEEE/ASME Journal of Microelectromechanical Systems*, 23(1):66–77, 2013.

127 P. Taheri-Tehrani, T. Su, A. Heidari, G. Jaramillo, C. Yang, S. Akhbari, H. Najar, S. Nitzan, D. Saito, L. Lin, and D. A. Horsley. Micro-Scale Diamond Hemispherical Resonator Gyroscope. In Solid-State Sensors, Actuators, and Microsystems Workshop (Hilton Head), Hilton Head Island, SC, USA, 2014.

128 L. D. Sorenson, X. Gao, and F. Ayazi. 3-D Micromachined Hemispherical Shell Resonators with Integrated Capacitive Transducers. In IEEE International Conference on Micro Electro Mechanical Systems (MEMS), Paris, France, 2012.

129 I. P. Prikhodko, S. A. Zotov, A. A. Trusov, and A. M. Shkel. Sub-Degree-per-Hour Silicon MEMS Rate Sensor with 1 Million Q-Factor. In Solid-State Sensors, Actuators and Microsystems Conference (TRANSDUCERS), Beijing, China, 2011.

130 D. Senkal, I. P. Prikhodko, A. A. Trusov, and A. M. Shkel. Micromachined 3-D Glass-Blown Wineglass Structures for Vibratory MEMS Applications. In Technologies for Future Micro-Nano Manufacturing Workshop, Napa Valley, CA, USA, 2011.

131 D. Senkal, M. J. Ahamed, and A. M. Shkel. Design and Modeling of Micro-Glassblown Inverted-Wineglass Structures. In IEEE International Symposium on Inertial Sensors and Systems, Laguna Beach, CA, USA, 2014.

132 J. Donea, S. Giuliani, and J. Halleux. An Arbitrary Lagrangian-Eulerian Finite Element Method for Transient Dynamic Fluid-Structure Interactions. *Computer Methods in Applied Mechanics and Engineering*, 33:689–723, 1982.

133 C. Hirt, A. Amsden, and J. Cook. An Arbitrary Lagrangian-Eulerian Computing Method for All Flow Speeds. *Journal of Computational Physics*, 253(14):227–253, 1974.

134 J. E. Shelby. *Introduction to Glass Science and Technology*. Royal Society of Chemistry, Cambridge, UK, 2005.

135 D. Senkal, M. J. Ahamed, M. H. A. Ardakani, S. Askari, and A. M. Shkel. Demonstration of 1 Million Q-Factor on Microglassblown Wineglass Resonators with Out-of-Plane Electrostatic Transduction. *IEEE/ASME Journal of Microelectromechanical Systems*, 24(1):29–37, 2015.

136 M. H. Asadian, Y. Wang, R. M. Noor, and A. M. Shkel. Design Space Exploration of Hemi-Toroidal Fused Quartz Shell Resonators. In IEEE International Symposium on Inertial Sensors and Systems, Naples, FL, USA, 2019.

137 V. F. Zhuravlev and D. M. Klimov. Hemispherical Resonator Gyro. Nauka, Moscow, Russia, 1985.

138 N. E. Egarmin and V. E. Yurin. Introduction to Theory of Vibratory Gyroscopes. Binom Publishing, Moscow, Russia, 1993.

139 S. Choi and J. H. Kim. Natural Frequency Split Estimation for Inextensional Vibration of Imperfect Hemispherical Shell. *Journal of Sound and Vibration*, 330(9):2094–2106, 2011.

140 S. Y. Choi, Y. H. Na, and J. H. Kim. Thermoelastic Damping of Inextensional Hemispherical Shell. *World Academy of Science, Engineering and Technology*, 56:198–203, 2009.

141 M. Eichler, B. Michel, P. Hennecke, M. Gabriel, and C. Klages. Low-Temperature Direct Bonding of Borosilicate, Fused Silica, and Functional

Coatings. In International Symposium on Semiconductor Wafer Bonding: Science, Technology, and Applications, volume 33, Las Vegas, NV, USA, 2010.

142 Y. Wang, M. H. Asadian, and A. M. Shkel. Modeling the Effect of Imperfections in Glassblown Micro-Wineglass Fused Quartz Resonators. *ASME Journal of Vibration and Acoustics*, 139(4):040909, 2017.

143 D. Senkal, M. J. Ahamed, A. A. Trusov, and A. M. Shkel. Electrostatic and Mechanical Characterization of 3-D Micro-Wineglass Resonators. *Sensors and Actuators A: Physical*, 215:150–154, 2014.

144 M. J. Ahamed, D. Senkal, A. A. Trusov, and A. M. Shkel. Deep NLD Plasma Etching of Fused Silica and Borosilicate Glass. In IEEE Sensors Conference, Baltimore, MD, USA, 2013.

145 D. Senkal, M. J. Ahamed, A. A. Trusov, and A. M. Shkel. Achieving Sub-Hz Frequency Symmetry in Micro-Glassblown Wineglass Resonators. *IEEE/ASME Journal of Microelectromechanical Systems*, 23(1):30–38, 2014.

146 D. Senkal, C. R. Raum, A. A. Trusov, and A. M. Shkel. Titania Silicate/Fused Quartz Flassblowing for 3-D Fabrication of Low Internal Loss Wineglass Micro-Structures. In Solid-State Sensors, Actuators, and Microsystems Workshop (Hilton Head), Hilton Head Island, SC, USA, 2012.

147 D. Senkal, M. J. Ahamed, A. A. Trusov, and A. M. Shkel. Adaptable Test-Bed for Characterization of Micro-Wineglass Resonators. In IEEE International Conference on Micro Electro Mechanical Systems (MEMS), Taipei, Taiwan, 2013.

148 D. Senkal, M. J. Ahamed, S. Askari, and A. M. Shkel. 1 Million Q-Factor Demonstrated on Micro-Glassblown Fused Silica Wineglass Resonators with Out-of-Plane Electrostatic Transduction. In Solid-State Sensors, Actuators and Microsystems Workshop (Hilton Head), Hilton Head Island, SC, USA, 2014.

149 A. M. Shkel and D. Senkal. Environmentally Robust Micro-Wineglass Gyroscope. *US Patent 9,429,428 B2*, 2016.

150 M. H. Asadian and A. M. Shkel, "*Fused quartz dual shell resonator,*" in IEEE Int. Symp. Inertial Sensors and Systems, April 2019.

151 M. H. Asadian, D. Wang, and A. M. Shkel, "*Microscale 3D Fused Quartz Dual-Shell Resonator Instrumented as a Rate Gyroscope,*" in Solid-State Sensors, Actuators, and Microsystems Workshop (Hilton Head), Hilton Head Island, SC, USA, Jun. 2020.

152 M. H. Asadian, Y. Wang, D. Wang, S. Peshin, and A. M. Shkel, "*Design of High-Q 3D Dual-Shell Resonators for High Shock Environments,*" in IEEE/ION Position Location and Navigation Symposium (PLANS), Portland, OR, USA, April 2020.

153 Khalil Najafi, Jae Yoong Cho, Sajal Singh, Tal Nagourney, Jong-Kwan Woo, Behrooz Shiari, Ali Darvishian, Guohong He, Doruk Senkal, Mohammad Asadian, Yusheng Wang, Danmeng Wang, Andrei M. Shkel "*Micromachined Gyroscopes Made From 3D Molded Fused-Silica Shell Resonators*", in IEEE/ION Position Location and Navigation Symposium (PLANS), Portland, OR, USA, April 2020.

Index

a

AC quadrature null   34, 60, 62
ADC. *See* Analog to Digital Converter
AFM. *See* atomic force microscope
AGC. *See* Amplitude Gain Control
Akheiser dissipation   16
ALD. *See* atomic layer deposition
ALE. *See* Arbitrary Lagrangian-Eulerian
Allan Variance   7–9, 40–41, 52–53, 60–61
amplitude control. *See* Amplitude Gain Control (AGC)
Amplitude Gain Control (AGC)   4, 24–28, 30–34, 60, 62
Analog to Digital Converter (ADC)   25
anchor losses   14–15, 48, 87, 94, 96
anchor loss mitigation   55–56
Angle Random Walk (ARW)   8, 40–41, 53, 60–61, 68, 136
angular gain factor   12, 18, 33, 62–63, 96, 98
angular velocity   5, 12, 18, 29, 60
aniso-damping   16–17, 19–20, 135
aniso-elasticity   16–20, 30, 135
anodic bonding   114
anti-phase motion   54–57, 97
Arbitrary Lagrangian-Eulerian (ALE)   92
ARW. *See* Angle Random Walk

atomic force microscope (AFM)   107
atomic layer deposition (ALD)   75

b

bandwidth   4, 27
bar resonators   5
batch-fabrication   123–24
BAW. *See* Bulk Acoustic Wave
bias instability   8, 136
blow-torch molding   73, 82
BMG. *See* Bulk Metallic Glass
bonding   43–44, 82, 102–4, 126, 128, 136
  glass frit   43
  low temperature eutectic   44
borosilicate glass   70, 79, 82, 124
Bulk Acoustic Wave (BAW)   47
Bulk Metallic Glass   71–72, 83–84

c

canonical variables   18–19, 30
capacitive gap   14, 48–49, 57–58, 83–84, 116–17, 123–24, 126, 128, 129, 136
cavity pressure   7, 93
central anchor   41, 47, 88–90, 105, 114–17, 136
chip-scale atomic devices   69
CO_2 laser   74–75, 80

Whole-Angle MEMS Gyroscopes: Challenges and Opportunities,
First Edition. Doruk Senkal and Andrei M. Shkel.
© 2020 The Institute of Electrical and Electronics Engineers, Inc. Published 2020 by John Wiley & Sons, Inc.

Coefficient of Thermal Expansion (CTE) 15, 75–76, 135
control strategies   9, 23–36
conventional drive   53–54
Coriolis forces   3–5, 11–12, 17–18, 20, 23, 123
Coriolis signal   23–24, 26, 34
Coriolis Vibratory Gyroscopes (CVGs)   3–4, 9, 11, 14, 18–19, 23, 26, 68, 123
cost, size, weight, and power (CSWaP)   3
CSWaP. *See* cost, size, weight, and power
CTE. *See* Coefficient of Thermal Expansion
CVGs. *See* Coriolis Vibratory Gyroscopes

d

DAC. *See* Digital to Analog Converter
damping   5, 12, 16–18, 29–30, 55
DC bias voltage   34, 51, 60, 119, 121
DC quadrature null   34–35
Deep Reactive Ion Etching   43–45
degenerate modes   3, 20, 33, 96, 98, 119–20
demodulation   23–24, 29
DFP. *See* Dual Foucault Pendulum
diamond   76, 78
Digital to Analog Converter (DAC)   25, 60
Disk Resonator Gyroscope (DRG)   41, 47, 134
dissipation mechanisms   13
Distributed Mass MEMS Gyroscope   47, 49, 51, 53
DRG. *See* Disk Resonator Gyroscope
drive mode   4, 24–25, 60
drive mode oscillator   24, 26–27
Dual Foucault Pendulum (DFP)   9, 47, 54–57, 59, 61, 63

e

EAM. *See* Electromechanical Amplitude Modulation

EDM. *See* electro-discharge machining
electrode assembly   49, 85, 111–12, 115
electrode configurations   119, 125
electrodes   33, 40, 48–50, 60, 67, 81–84, 111–14, 116–17, 119, 123–25, 128, 135
  assembled   81, 111–15
  discrete   49–50, 84
  in-plane   10, 69, 111, 114
  integrated   83–84, 116–18, 126
  metal   135
  out-of-plane   10, 111, 123–25, 129, 135
  radial   123–24
  silicon   82, 85
  temporary   111
electro-discharge machining (EDM)   69, 73, 76
Electromechanical Amplitude Modulation (EAM)   50
electrostatic tuning   51–52, 60, 134
elliptical orbit   18–19
energy decay time constant   12, 16, 20, 60
energy loss   13
EpiSeal. *See* Epitaxial Silicon Encapsulation Process
Epitaxial Silicon Encapsulation Process   14, 41, 44–45, 49–50

f

FEA. *See* Finite Element Analysis
FEM. *See* finite element method
fiber optic gyro (FOG)   69
Finite Element Analysis (FEA)   15, 57, 90–96
finite element method (FEM). *See* Finite Element Analysis (FEA)
FOG. *See* fiber optic gyro
force and torque balance   55–56
forcer   34, 49–52, 59–60, 118, 123, 125, 129, 135

force-to-rebalance (FTR)   5, 24, 27–29, 31–32, 35–36, 52–53, 61–62
Foucault Pendulum   11–12, 55–56, 96
FRB. *See* force-to-rebalance (FTR)
frequency split   16, 41, 50–51, 101, 120–21, 129, 130, 134
frequency symmetry   20, 98, 100, 121
front-end amplifier   25–26
FS. *See* fused silica
FTR. *See* force-to-rebalance   27, 53
fused silica (FS)   9, 15, 21, 67, 73, 87–88, 92, 102, 104–5, 109–11, 124–25, 133, 136
fused silica resonator   68, 123, 129, 131

g

getter pump   59
glass   70–72, 76, 83, 85, 92, 112, 116, 124
glassblowing   70, 88–89, 92–93, 95, 103–10, 115–16
glassblown structure   88, 105
gyroscope   3–9, 12, 14–20, 24, 26–30, 32, 34, 39–41, 43, 47, 49, 55–57, 59–61, 68, 133–34
  automotive   36
  axisymmetric   96
  cylindrical   76
  degenerate mode   5, 35–36, 55
  disk   6, 47
  force-feedback   27
  lumped mass   9, 47, 96
  micro-wineglass   67
  mode-match   27–28, 36
  mode-mismatched   35
  nondegenerate mode   3–4, 24
  open-loop   26
  opto-mechanical   81
  rate   24–25, 27, 30, 40–41, 52, 60
  rate-integrating   5, 11, 18, 20, 29, 98, 121
  ring   47–48

whole-angle   5–6, 9, 11, 20–21, 23, 32, 35, 41, 67, 133, 135
wineglass   9, 55, 96, 135

h

Hemispherical Resonator Gyroscope (HRG)   67–68, 87, 100
hemitoroid   102
hermetic seal   43–44
HF. *See* Hydrofluoric Acid
HRG. *See* Hemispherical Resonator Gyroscope
Hydrofluoric Acid (HF)   74, 76, 112

i

in-run bias stability   26, 40–41, 53, 60
instrumentation amplifier   59
integration time   8–9, 53

l

laser ablation   83, 95, 105–7, 134
Laser Doppler Vibrometry (LDV)   73, 78, 118, 121
LDV. *See* Laser Doppler Vibrometry
low CTE materials   15
low-outgassing ceramic   58–59
Low Pressure Chemical Vapor Deposition (LPCVD)   44
LPCVD. *See* Low Pressure Chemical Vapor Deposition
lumped mass   54–55, 57, 59, 61, 63

m

macro-scale devices   100
macro-scale shell resonator gyroscopes   9, 67, 87
mechanical trimming   133–34
micro-glassblowing   69, 72, 80, 87, 89–90, 101–4, 109, 111, 115, 136
  process   71, 88, 92–94, 114, 121, 124
micro-glassblown
  shells   106–7, 109
  structures   87, 101–2

micro-glassblown (cont'd)
   wineglass resonators 10, 88–118, 122, 124, 126, 128, 129, 135
Micro-rate Integrating Gyroscope (MRIG) 6
micro-shell 74, 79, 84
   resonators 69–70, 79–85
   structures 21, 74, 79, 81–82, 85
micro-wineglass 5–6, 67, 73, 80–81, 91–92, 111, 113, 117, 120
mode shapes 118, 121
MRIG. *See* Micro-rate Integrating Gyroscope

n

noise 6–8, 41

o

open-loop mechanization 4, 24–29, 36
optical profilometry 74
oscillation, drive mode 4, 24

p

parallel plate 39, 57–58
parametric drive 23, 31–34, 48, 50, 53–54
parametric pumping effect 33
pattern angle 18–20, 30–32, 35, 40, 53, 62–63
Perfectly Matched Layers (PML) 14–15, 96
Phase Locked Loop (PLL) 24–28, 30–31, 34, 60, 62
PID. *See* Proportional Integral Derivative
PLL. *See* Phase Locked Loop
PML. *See* Perfectly Matched Layer
Poached Egg micro-shell resonators 78–79, 85
polycrystalline diamond 21, 75–76, 79, 83, 133
PolySi. *See* polysilicon
polysilicon 40, 83–84, 103

precession pattern 18, 20, 30, 33–35, 52
primary modes 5, 14–16, 33, 134
principal axes of damping 17
principal axes of elasticity 16–17
proof mass 11–12, 39, 55–57, 96
Proportional Integral Derivative (PID) 26–27, 29, 31–32, 34

q

Q-factor 12–13, 16, 21, 40–41, 48, 51, 60, 67–68, 73, 76, 78, 87, 126, 129–31
QMG. *See* Quadruple Mass Gyroscope
quadrature 19–20, 23, 29, 35
quadrature error 19–20, 23, 30–31
quadrature null loop 28–32, 34, 60, 62
quadrature signal 23–24, 26, 29, 34
Quadruple Mass Gyroscope (QMG) 40, 55
quantization 8, 25

r

ratchet mechanism 113
Rate Random Walk (RRW) 8
rate table 59–60, 63
recrystallization 108–9
resonance frequency 12–13, 16, 33–34, 50, 56, 101
resonator 12–13, 59–60, 67, 69, 83, 95, 97–101, 111, 113, 117, 119, 123, 125–26
RIG. *See* gyroscope, rate-integrating
ring gyroscopes 40–41, 47
ring laser gyro (RLG) 69
RLG. *See* ring laser gyro
RRW. *See* Rate Random Walk

s

scale factor 7, 26, 28, 60, 135
scale factor errors 6–7
self-calibration 4, 9, 134–35
sense mode 4, 24–25, 27–29, 39, 60
shuttle assemblies 39, 56–58

silicon 15, 21, 44–45, 47–49, 55, 70, 74–75, 82, 84, 87, 116
silicon mold 74–75, 84
silicon vias 45–46
SiO_2 44–45, 76, 78, 109
softening point 70, 88
spherical resonator 114–15
spherical shells 69, 71, 88–89, 95, 105
spokes 41, 47, 49, 134
sputtering 78, 116–17
standing wave pattern 32–33, 97
stem 84, 92, 95–96, 113, 128
structural imperfections 19–20, 133–34
  effects of 9, 18–20
structural symmetry 131, 133
substrate 14–15, 48, 70, 83–84, 88–89, 92, 94, 103, 115–17
surface losses 13, 16, 107
surface roughness 21, 76, 88, 101, 106–8
surface tension forces 70, 88, 101–2, 107
symmetry, damping 20

t

TED. *See* thermoelastic dissipation
thermoelastic dissipation (TED) 15, 87
Titania Silicate Glass (TSG) 9, 21, 87, 89, 102–3, 108–10
Toroidal Ring Gyroscope (TRG) 9, 47–49, 51–53
TRG. *See* Toroidal Ring Gyroscope
TSG. *See* Titania Silicate Glass
tuning fork gyroscope 55–56

u

ULE. *See* Ultra Low Expansion Glass
ULE TSG. *See* Ultra Low Expansion Titania Silicate Glass
Ultra Low Expansion Glass (ULE) 78–79
Ultra Low Expansion Titania Silicate Glass 21, 78–79, 85, 87, 110

v

vacuum chamber 59
vector drive 32, 34, 53
vibratory modes 3–4, 17
vibratory structure 13, 15, 41, 48
Virtual Carouseling 35, 60, 135
viscous damping 13–14, 60, 129

w

wafer 43, 70, 74, 81, 103, 111–12, 115, 124–26, 128
whole-angle mechanization 4–5, 29, 31, 33, 35–36, 41, 55, 60, 135
wineglass 5, 69, 91, 93, 95, 100–101, 105–6, 112–13, 116, 118, 128
wineglass modes 40–41, 49–51, 53, 55, 81, 94, 96–98, 100, 119–21, 123, 125
wineglass resonators 68, 87, 97–98, 101–2, 115–16, 118, 121, 123–24
wineglass structure 88–89, 95–96, 103, 106, 115–16, 123

x

XeF_2 etching 74–75, 79, 83, 116

IEEE Press Series on Sensors

Series Editor: Vladimir Lumelsky, Professor Emeritus, Mechanical Engineering, University of Wisconsin-Madison

Sensing phenomena and sensing technology is perhaps the most common thread that connects just about all areas of technology, as well as technology with medical and biological sciences. Until the year 2000, IEEE had no journal or transactions or a society or council devoted to the topic of sensors. It is thus no surprise that the IEEE Sensors Journal launched by the newly-minted IEEE Sensors Council in 2000 (with this Series Editor as founding Editor-in-Chief) turned out to be so successful, both in quantity (from 460 to 10,000 pages a year in the span 2001–2016) and quality (today one of the very top in the field). The very existence of the Journal, its owner, IEEE Sensors Council, and its flagship IEEE SENSORS Conference, have stimulated research efforts in the sensing field around the world. The same philosophy that made this happen is brought to bear with the book series.

Magnetic Sensors for Biomedical Applications
Hadi Heidari, Vahid Nabaei
Smart Sensors for Environmental and Medical Applications
Hamida Hallil, Hadi Heidari
Whole-Angle MEMS Gyroscopes: Challenges, and Opportunities
Doruk Senkal and Andrei M. Shkel.

Printed and bound by CPI Group (UK) Ltd, Croydon, CR0 4YY
19/04/2023
03212131-0003